VOLUME ONE HUNDRED AND FORTY FIVE

ADVANCES IN CANCER RESEARCH

GPCR Signaling in Cancer

VOLUME ONE HUNDRED AND FORTY FIVE

ADVANCES IN CANCER RESEARCH

GPCR Signaling in Cancer

Edited by

ARUN K. SHUKLA

Department of Biological Sciences and Bioengineering, Indian Institute of Technology, Kanpur, India

ELSEVIER

ACADEMIC PRESS

An imprint of Elsevier

Academic Press is an imprint of Elsevier
50 Hampshire Street, 5th Floor, Cambridge, MA 02139, United States
525 B Street, Suite 1650, San Diego, CA 92101, United States
The Boulevard, Langford Lane, Kidlington, Oxford OX5 1GB, United Kingdom
125 London Wall, London, EC2Y 5AS, United Kingdom

First edition 2020

ISBN: 978-0-12-820230-2
ISSN: 0065-230X

For information on all Academic Press publications
visit our website at https://www.elsevier.com/books-and-journals

Publisher: Zoe Kruze
Editorial Project Manager: Leticia M. Lima
Production Project Manager: Vijayaraj Purushothaman
Cover Designer: Victoria Pearson

Typeset by SPi Global, India

Contents

Contributors

Anthony Ashton
Pennsylvania Cancer and Regenerative Medicine Research Center, Baruch S. Blumberg Institute, Pennsylvania Biotechnology Center, Wynnewood, PA, United States; Division of Perinatal Research, Royal North Shore Hospital, St. Leonards, NSW, Australia

Martin Augsten
amcure GmbH, Eggenstein-Leopoldshafen, Germany

Andy Chevigné
Department of Infection and Immunity, Immuno-Pharmacology and Interactomics, Luxembourg Institute of Health (LIH), Esch-sur-Alzette, Luxembourg

Hemlata Dwivedi-Agnihotri
Department of Biological Sciences and Bioengineering, Indian Institute of Technology, Kanpur, India

James J. Hoy
LCMB Center for Cancer Research, National Cancer Institute, Bethesda, MD, United States

Xuanmao Jiao
Pennsylvania Cancer and Regenerative Medicine Research Center, Baruch S. Blumberg Institute, Pennsylvania Biotechnology Center, Wynnewood, PA, United States

Georgios Kallifatidis
Georgia Cancer Center; Department of Biological Sciences, Augusta University; Research Service, Charlie Norwood Veterans Administration Medical Center, Augusta, GA, United States

Andrew V. Kossenkov
Wistar Institute, Philadelphia, PA, United States

Bal L. Lokeshwar
Georgia Cancer Center, Augusta University; Research Service, Charlie Norwood Veterans Administration Medical Center, Augusta, GA, United States

Max Meyrath
Department of Infection and Immunity, Immuno-Pharmacology and Interactomics, Luxembourg Institute of Health (LIH), Esch-sur-Alzette, Luxembourg

Arne Östman
Department of Oncology-Pathology, Karolinska Institutet, Stockholm, Sweden

Kishan Patel
Pennsylvania Cancer and Regenerative Medicine Research Center, Baruch S. Blumberg Institute, Pennsylvania Biotechnology Center, Wynnewood, PA; Xavier University School of Medicine, Woodbury, NY, United States

Richard G. Pestell
Pennsylvania Cancer and Regenerative Medicine Research Center, Baruch S. Blumberg Institute, Pennsylvania Biotechnology Center, Wynnewood; Wistar Institute, Philadelphia, PA; Xavier University School of Medicine, Woodbury, NY, United States

Arun K. Shukla
Department of Biological Sciences and Bioengineering, Indian Institute of Technology, Kanpur, India

Elin Sjöberg
Department of Oncology-Pathology, Karolinska Institutet, Stockholm, Sweden

Surinder M. Soond
Institute of Molecular Medicine, Sechenov First Moscow State Medical University, Moscow, Russian Federation

Martyna Szpakowska
Department of Infection and Immunity, Immuno-Pharmacology and Interactomics, Luxembourg Institute of Health (LIH), Esch-sur-Alzette, Luxembourg

Chandan Upadhyaya
Pennsylvania Cancer and Regenerative Medicine Research Center, Baruch S. Blumberg Institute, Pennsylvania Biotechnology Center, Wynnewood, PA; Xavier University School of Medicine, Woodbury, NY, United States

Andrey A. Zamyatnin Jr.
Institute of Molecular Medicine, Sechenov First Moscow State Medical University; Belozersky Institute of Physico-Chemical Biology, Lomonosov Moscow State University, Moscow, Russian Federation

CHAPTER ONE

Atypical chemokine receptors in tumor cell growth and metastasis

Bal L. Lokeshwar[a,b,*], Georgios Kallifatidis[a,b,c], James J. Hoy[d]
[a]Georgia Cancer Center, Augusta University, Augusta, GA, United States
[b]Research Service, Charlie Norwood Veterans Administration Medical Center, Augusta, GA, United States
[c]Department of Biological Sciences, Augusta University, Augusta, GA, United States
[d]LCMB Center for Cancer Research, National Cancer Institute, Bethesda, MD, United States
[*]Corresponding author: e-mail address: blokeshwar@augusta.edu

Contents

Abstract

Atypical chemokine receptors (ACKRs) are seven-transmembrane cell surface protein receptors expressed in immune cells, normal mesenchymal cells, and several tumor cells. As of this writing, six ACKRs have been characterized by diverse activities. They bind both cysteine-cysteine (CC) type and cysteine-X-cysteine (CXC)-type chemokines, either alone, or together with a ligand bound-functional G-protein coupled (typical) chemokine receptor. The major structural difference between ACKRs and typical chemokine receptors is the substituted DRYLAIV amino acid motif in the second intracellular loop of the ACKR. Due to this substitution, these receptors cannot bind Gαi-type G-proteins responsible for intracellular calcium mobilization and cellular chemotaxis. Although initially characterized as non-signaling transmembrane receptors (decoy receptors) that attenuate ligand-induced signaling by GPCRs, studies of all ACKRs have shown ligand-independent and ligand-dependent transmembrane signaling in both non-tumor and tumor cells. The precise function and mechanism of the differential expression of ACKRs in many tumors are not understood well. The use of antagonists of ACKRs ligands has shown limited antitumor potential; however, depleting ACKR expression resulted in a reduction in experimental tumor growth and metastasis. The ACKRs represent a unique class of transmembrane signaling proteins that regulate growth,

Advances in Cancer Research, Volume 145
ISSN 0065-230X
https://doi.org/10.1016/bs.acr.2019.12.002

survival, and metastatic processes in tumor cells, affecting multiple pathways of tumor growth. Therefore, closer investigations of ACKRs have a high potential for identifying therapeutics which affect the intracellular signaling, preferentially via the ligand-independent mechanism.

Abbreviations

ACKR	atypical chemokine receptors
AR	androgen receptor
ARRB	β-arrestin
CXCL	cysteine-X-cysteine-motif chemokine ligand
CXCR	cysteine-X-cysteine chemokine receptor
DRYLAIV	amino acids motif that binds G-protein α_i
E2	estradiol
EGF	epidermal growth factor
EGFR	EGF receptor
ERα	estrogen receptor-α
GPCR	G-protein coupled (chemokine) receptors

1. Introduction

Chemokine receptors are members of a large family of protein-ligand binding cell surface receptors and contain seven-transmembrane domains. Chemokine receptors play a key role in immune cell functions, and it is becoming increasingly evident that they also contribute to cancer cell survival and metastasis. They are classified as *typical* (G-protein coupled receptors) and *atypical* chemokine receptors. As the GPCRs have been abundantly described in this volume elsewhere, this article will detail the role of ACKR, mainly in cancers. ACKRs are so named and approved by the Human Genome Organization Gene Nomenclature Committee and Nomenclature Committee of the International Union of Pharmacology in 2014 (Bachelerie, Ben-Baruch, et al., 2014; Bachelerie, Graham, et al., 2014). Accordingly, the official name of all ACKRs is referred to so, but they continue to be referenced based on their name at the time of their discovery. The main structural distinction between GPCR and ACKR is the modified DRYLAIV motif in the second intracellular loop of the seven-transmembrane receptor protein, where the modification is specific to each of the five ACKRs (Fig. 1). Due to these amino acid substitutions, ACKRs do not bind the Gαi subunit of the G-proteins and they functionally do not activate intracellular calcium (Ca^{2+}) mobilization and cellular motility.

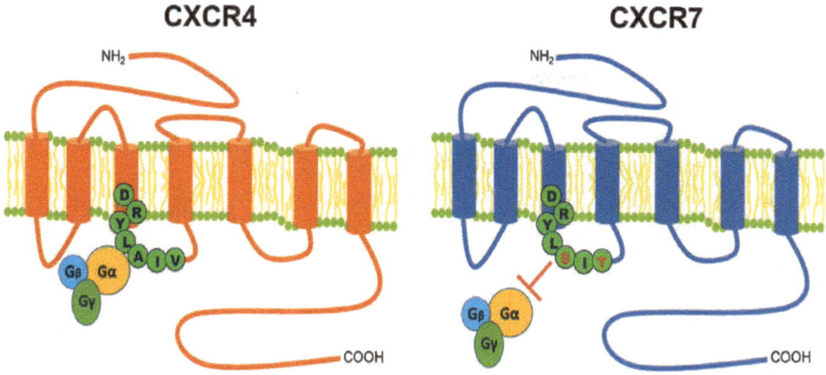

Fig. 1 An illustration of the secondary structures of GPCR and ACKR: difference in the secondary amino acid structures (DRY motif) between a typical GPCR (e.g., CXCR4) and ACKR (e.g., CXCR7). Alteration in the DRYLAIV motif in the second intracellular loop of the ACKR protein negates its interaction with G-protein such as Gαi resulting in the lack of Ca^{2+} mobilization and ligand-induced motility.

As summarized in Table 1, the extracellular domain of each of the six ACKRs binds a variety of ligands. Ligand bound ACKR often dimerizes with its co-receptor; usually, a GPCR, internalizes the ligand and help elicit various cellular activities.

Following the discovery of ACKR1, also known as the Duffy antigen, about 65 years ago four, more ACKRs have been identified to date. Besides, a putative ACKR6 has been identified in human and mouse genome databases (Bachelerie, Ben-Baruch, et al., 2014). As for the HUGO Gene Nomenclature Committee database, this gene encodes a member of a family of membrane-associated phosphatidylinositol transfer domain-containing proteins (PITPNM3, or PITPNM family member3). The PITPNM3 (Lin et al., 2016) protein is homologous to a Drosophila protein that is implicated in the visual transduction pathway in flies (Han et al., 2019; Hong, Shen, Xie, & Ruan, 2012).

The ACKR1 was the first one to be characterized as a scavenger receptor for at least 20 CC or CXC chemokines and had broad specificity and affinities for various ligands. It lacks the entire Gαi binding domain DRYLAIV consensus motif in its second intracellular loop and has less than 25% homology with known GPCRs (Chaudhuri et al., 1994; Neote, DiGregorio, Mak, Horuk, & Schall, 1993; Neote, Mak, Kolakowski, & Schall, 1994). All ACKR1 ligands have a functional role in inflammation. However, it binds a unique antigen belonging to the tetraspanin family antigen CD82 that was known as a tumor suppressor for prostate cancer (Bandyopadhyay et al., 2006). Considering the important role of CD82 in prostate cancer cell

Table 1 A summary of various ACKRs, their cognate ligands and major functions.

IUP[a] name	Alternate names	Ligands	Expression	Common functions
ACKR1	Duffy antigen, DRAC, CD234	CCLs (2,7,8,13,14,16,17,22); CXCLs (1,5,6,7,8,9,11,13)	Erythrocytes, endothelial cells, tumor vasculature cells and neurons. Pancreatic cancer cells	Ligand internalization, presentation, sink and reservoir, port of entry for *P. vivax*. Mostly tumor suppressive function by inhibiting vasculogenesis
ACKR2	D6, CCR9, CCR10, CCBP2, CMKBR9	CCLs (2,3,3L1,4,5,7,8,11, 13,14,17,22)	Macrophage subsets, lymphatic endothelial cells, B-cells, and tumor endothelial cells	CC inflammatory chemokine scavenger. Tumor suppressor. Negatively associated with tumor metastasis
ACKR3	RDC1, CXCR7, CMKOR1	CXCL11 and CXCL12	Mesenchymal and hematopoietic cells, neurons, glia, tumor cells and some normal epithelial cells, tumor endothelial cells	Ligand scavenging, chemotaxis, mitogenic signaling. Tumor metastasis
ACKR4	CCRL1, CCX-CKR, CCR1	CCL19, CCL21, CCL25 and CXCL13	Thymic epithelial cells, bronchial epithelial cells, and keratinocytes	Ligand scavenging. Tumor suppressor function in some tumor systems such as hepatocellular carcinoma
ACKR5	CCRL2	CCL2, chemerin	Hematopoietic cells, astrocytes, microglia and lung epithelial cells. Breast, colon, prostate and glioblastoma tumor cells	Trafficking of CMKRL1 expressing leucocytes
ACKR6	PITPNM3	CCL18	Tumor-associated macrophages, breast cancer, hepatocellular and pancreatic ductal carcinoma	Receptor for CCL18, cell adhesion, alternation of VCAM1 function

[a]IUP: International Union of Pharmacology.

migration and silencing in aggressive cancers, ACKR1 structure and functions are reviewed in this chapter. The ACKR2, formerly known as the D6 receptor discovered in 1997, is a scavenging receptor for at least a dozen CC chemokines (Bonini, Anderson, & Steiner, 1997; Bonini & Steiner, 1997; Nibbs, Wylie, Pragnell, & Graham, 1997). It was also the first atypical chemokine receptor that was shown to act as a decoy and a scavenging receptor (Fra et al., 2003). The ACKR3 is another seven-transmembrane receptor that was also discovered accidentally and named an orphan receptor. Cloned from a canine cDNA library and identified as an orphan receptor (Libert, Parmentier, Lefort, Dumont, & Vassart, 1990), it was initially labeled as a receptor for vasoactive intestinal peptide-1 (VIP-1) and later disputed and alternatively termed as calcitonin-gene related peptide (CGRP) receptor. However, it was not until 2005 that the stromal-derived chemokine-1α, was identified as a ligand in human T-lymphocytes (Balabanian et al., 2005; Kapas & Clark, 1995; Nagata et al., 1992; Sreedharan, Robichon, Peterson, & Goetzl, 1991). As a receptor for SDF-1α, also known as CXC chemokine ligand-12 (CXCL12), RDC1/ACKR3 was relabeled as CXC-chemokine receptor 7 (CXCR7) and a year later Burns et al. reported CXCR7 binds to an iTAC-1 antigen, CXCL11, and was a co-receptor for CXCL12 along with CXCR4 (Balabanian et al., 2005; Burns et al., 2006). ACKR3 has been shown to be one of the most relevant ACKRs in tumor growth and metastasis. Many human tumor cells of epithelial origin as well as lymphomas overexpress ACKR3 (D'Alterio et al., 2010; Hattermann et al., 2010; Heinrich, Lee, Lu, Lowy, & Kim, 2012; Iwakiri et al., 2009; Kollmar et al., 2010; Liberman et al., 2012; Maksym et al., 2009; Meijer, Ogink, & Roos, 2008; Miao et al., 2007; Monnier et al., 2012; Salmaggi et al., 2009; Sun et al., 2010; Wang et al., 2008; Xue et al., 2012; Zheng et al., 2010). The mesenchymal tissue derived tumors, such as rhabdomyosarcoma, neuroblastoma, lymphoma, among others, also express ACKR3. Similarly, cell lines and tissues of epithelial tumors such as prostate, bladder, colorectal, stomach, pancreas, and lung cancers also express ACKR3 (Burns et al., 2006; D'Alterio et al., 2010; Goldmann et al., 2008; Grymula et al., 2010; Hattermann et al., 2010; Heinrich et al., 2012; Iwakiri et al., 2009; Kollmar et al., 2010; Lee, Song, Yun, Jo, & Kim, 2014; Liberman et al., 2012; Maksym et al., 2009; Meijer et al., 2008; Miao et al., 2007; Salmaggi et al., 2009; Singh & Lokeshwar, 2011; Sun et al., 2010; Tarnowski et al., 2010; Wurth et al., 2011; Xue et al., 2012). The following is a summary of various functions identified to date in tumor growth, vasculogenesis, and metastasis.

2. Specific ACKRs

2.1 ACKR1/DARC

The atypical chemokine receptor 1 (ACKR1) functions as a chemokine transporter (Borroni, Savino, Bonecchi, & Locati, 2018; Nibbs & Graham, 2013). When chemokines bind to ACKR1 expressed on polarized cells, they become internalized, transported across the cytoplasm and released on the other side of the cell for presentation to leucocytes. This process of transcytosis may promote the transmigration of leucocytes. Several studies demonstrated that ACKR1 expressing endothelial cells mediate the internalization of cytokines through the process of transcytosis. ACKR1 binds several cytokines (more than 20 different CC and CXC chemokines) (Atzler & de Winther, 2018; Salazar & Zabel, 2019) and is expressed in neurons of the cerebellum, red blood cells, and endothelial cells of venules and small veins (Atzler & de Winther, 2018; Chaudhuri et al., 1994; Rot, 2005).

Preclinical modes and studies performed in human subjects demonstrated that ACKR1 expression in erythrocytes negatively regulates the bioavailability of chemokines. ACKR1 functions as a cytokine scavenger and individuals lacking ACKR1 are characterized by an increased concentration of cytokines in the blood (Howes et al., 2011; Shen, Schuster, Stringer, Waltz, & Lentsch, 2006).

Recent evidence suggests that its expression inversely correlates with tumor progression in several cancer models. It was initially studied in the context of prostate cancer. While ACKR1 knockout mice exhibited increased tumor growth rates, studies on African American men could not confirm a correlation between ACKR1 and prostate cancer incidence (Elson et al., 2011; Nemesure, Wu, Hennis, & Leske, 2015; Shen et al., 2006). Tumors such as pancreatic cancer express ACKR1 on their cell surface, where it inhibits tumor cell proliferation (Maeda et al., 2017). Similarly, when expressed on lung and breast cancer cell lines, it interferes with tumor growth by inhibiting tumor angiogenesis and metastasis (Addison, Belperio, Burdick, & Strieter, 2004; Wang et al., 2006). Further studies have confirmed a protective role of ACKR1 when expressed on endothelial cells. Overexpression of ACKR1 in endothelial cells reduced angiogenesis and metastasis of melanoma cells (Horton, Yu, Zaja-Milatovic, Strieter, & Richmond, 2007) by neutralization of CXCR2 ligands, and CXCR2 mediated angiogenesis.

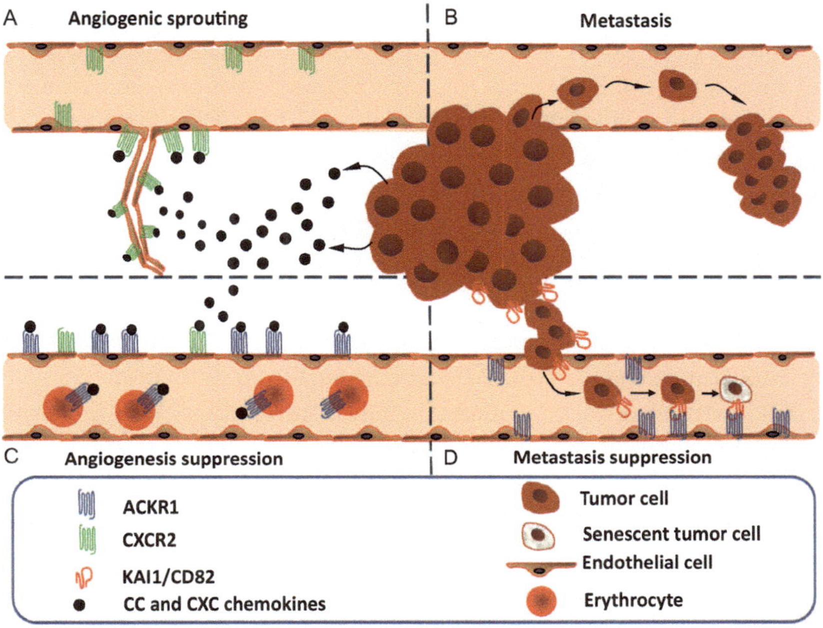

Fig. 2 Role of ACKR1 in cancer. ACKR1 functions as a tumor suppressor in several cancers. It is expressed in erythrocytes and endothelial cells, where it binds angiogenic cytokines and regulates their bioavailability. ACKR1 is also expressed on the endothelial cells of tumor-associated blood vessels, where it interacts with CD82 expressing tumor cells mediating their senescence.

However, ACKR1 was also shown to inhibit metastasis in cytokine-independent mechanisms. As mentioned above ACKR1 is expressed on endothelial cells and evidence suggests that it functions as a receptor for the tumor metastases suppressor protein KAI1/CD82. CD82 belongs to the family of tetraspanins and when expressed on tumor cells, it suppresses metastatic potential. When CD82 expressing tumor cells dislodge from the primary tumor and intra-vasate into vessels, they interact with ACKR1 expressing endothelial cells. The ACKR1-CD82 interaction induces signaling that upregulates p21 followed by tumor cell senescence, thus preventing metastasis of tumors cells (Iiizumi, Bandyopadhyay, & Watabe, 2007).

Collectively, current research suggests that ACKR1 functions as a tumor suppressor in several cancers (Fig. 2).

2.2 ACKR2/D6

The atypical chemokine receptor 2 (ACKR2) is expressed in several tissues including the placenta, lymphatic endothelial cells and certain leucocyte populations (Graham & Locati, 2013; Martinez de la Torre et al., 2007; McKimmie et al., 2008). It constitutively traffics between the cell membrane and endosomes. ACKR2 functions as a CC inflammatory chemokine scavenger, and extracellular chemokines bound to ACKR2 become internalized and delivered to lysosomes for degradation (Borroni et al., 2018; Nibbs & Graham, 2013). Furthermore, ACKR2 interacts with β-arrestins and drives downstream signaling promoting ACKR2 trafficking to the cell membrane (Nibbs & Graham, 2013).

The role of ACKR2 in cancer was first studied in an inflammation-driven model of skin carcinogenesis (Nibbs et al., 2007), where ACKR2 depletion promoted TPA/DMBA-mediated tumor formation. Similarly, ACKR2 knockout mice exhibited increased susceptibility to tumor development in models of inflammation-induced colon cancer (Vetrano et al., 2010). Consistently, ACKR2 expression seems to be decreased in lymphatic endothelial cells and mononuclear cells of a colon adenocarcinoma cohort compared to corresponding healthy controls (Langenes et al., 2013). In contrast, while ACKR2 depletion resulted in increased macrophage infiltration in diethyl nitrosamine induced liver tumors, it did not affect tumorigenesis (Schneider et al., 2012).

A protective role of ACKR2 is also evident in vascular tumors such as the Kaposi sarcoma (McKimmie et al., 2013; Savino et al., 2014). Consistently, ACKR2 seems to be downregulated in an ERK signaling mediated fashion in aggressive forms of this cancer (Savino et al., 2014). ACKR2 is also present in breast cancer, where it inversely correlates with lymph node metastasis (Wu et al., 2008). In some instances, ACKR2 becomes downregulated during the progression of breast cancer which is also accompanied by depletion of ACKR1 and ACKR4 and results in a worse outcome of the disease (Zeng et al., 2011, 2014). Single-nucleotide polymorphisms in the ACKR2 coding sequence are associated with increased chemokine scavenging potential and correlated with increased relapse-free survival (Yang et al., 2013; Yu, Wang, Yang, Zeng, & Shao, 2015).

In summary, ACKR2-mediated uptake and degradation of inflammatory cytokines interfere with the recruitment of pro-tumorigenic lymphocytes and negatively regulates inflammation-induced tumorigenesis in several cancers (Fig. 3).

2.3 ACKR3/RDC1/CXCR7

After several misidentifications of the ACKR3 putative ligands, it was termed an orphan receptor until the mouse ACKR3/RDC1 was shown to

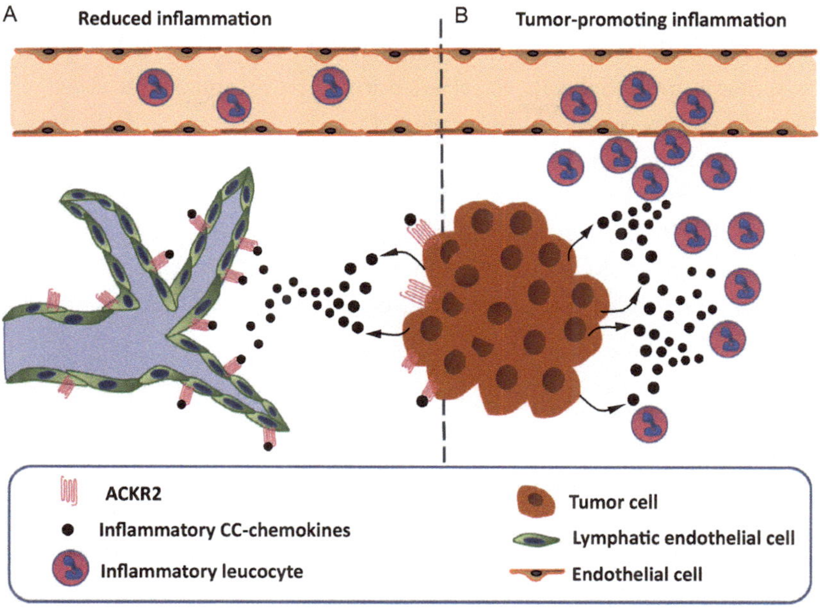

Fig. 3 Role of ACKR2 in cancer. ACKR2 is expressed in lymphatic endothelial cells; however its expression is decreased in tumor samples compared to the healthy tissue, which is associated with an increase in lymphatic vessel density (Langenes et al., 2013; Massara, Bonavita, Mantovani, Locati, & Bonecchi, 2016). It has a protective role in tumorigenesis by scavenging inflammatory cytokines such as CCL2, CCL3 and CCL5 interfering with macrophage/lymphocyte infiltration and angiogenesis (Massara et al., 2016).

localize on chromosome 1, along with sequence similarity to CXCR4 and CXCR2 which led to the hypothesis that RDC1 is a chemokine receptor (Heesen et al., 1998). Mouse genetic studies using ACKR3 knockout showed that homozygous ACKR3 knockout mice (ACKR3$^{-/-}$) die either perinatal or in utero due to cardiac septal defect. However, they observed normal mouse embryogenesis until day 17.5 in ACKR3$^{-/-}$ mice (Sierro et al., 2007). The homozygous ACKR3 knockout pups died shortly after birth due to ventricular septal defect and semilunar heart valve malformation (Sierro et al., 2007). Later it was reported that ACKR3 regulates availability and signaling of adrenomedullin, a mitogenic peptide hormone that regulates cardiovascular development (Klein et al., 2014). We did not find reports on the role of ACKR3 on human cardiac development.

2.3.1 Normal tissue distribution of ACKR3

A variety of normal and pathologic tissues express ACKR3. Using cDNA expression analysis (in situ hybridization, real-time-quantitative PCR,

Northern blotting and RNA-seq, etc.) ACKR3 was detected in mouse heart, kidney, spleen, lung and brain. ACKR3 is expressed predominantly in vascular endothelial cells, myeloid-derived endothelial cells and stromal cells such as fibroblasts and hyperplastic tissues (Su et al., 2002). ACKR3 is expressed in most mesenchymal stromal cells and in brain cells, such as glial cells, astrocytes, and in neuronal cells (Banisadr, Podojil, Miller, & Miller, 2016; Puchert et al., 2017; Virgintino et al., 2013). In leukocytes, the expression of ACKR3 protein is unclear, although some expression analyses report the detection of ACKR3 mRNA (Berahovich et al., 2010).

2.3.2 ACKR3 expression in tumor tissues and cells

Following discovery and reclassification of RDC1 as CXCR7 (ACKR3), its expression was detected and found elevated in many human epithelial cancer (carcinoma) cells. It was first reported in prostate tumors and established tumor cells using cDNA based gene expression analysis (Sun et al., 2003). Subsequently, ACKR3 is identified in most of the epithelial tumors (D'Alterio et al., 2010; Freitas et al., 2014; Gahan et al., 2012; Gosalbez et al., 2014; Grymula et al., 2010; Hattermann et al., 2010; Heinrich et al., 2012; Iwakiri et al., 2009; Kollmar et al., 2010; Liberman et al., 2012; Luker, Steele, Mihalko, Ray, & Luker, 2010; Maksym et al., 2009; Meijer et al., 2008; Salmaggi et al., 2009; Wang et al., 2008; Wurth et al., 2011; Zheng et al., 2010). Most endothelial cells express both ACKR3 and CXCR4, its CXCL12 binding partner (Bachelerie, Ben-Baruch, et al., 2014). However, many tumor cells, such as prostate and breast cancer cells, express either ACKR3 or CXCR4, seldom both (Burns et al., 2006; Salazar et al., 2014; Singh & Lokeshwar, 2011). When co-expressed in tumors, the levels of CXCR7 mRNA and protein are relatively high as compared to that of CXCR4 (Gahan et al., 2012; Mitchell et al., 2019; Monnier et al., 2012; Salazar et al., 2014; Yates et al., 2013).

2.3.3 Structure and function of ACKR3

ACKR3 is a highly conserved gene among mammals. As reported by Heesen et al. (1998), the cDNA sequence of mouse RDC1 had a size of 1861 nucleotide base pairs (bp). They identified an open reading frame between nucleotides 89 and 1177 that codes for 362 amino acids. The hydropathy plot of the sequenced protein showed the features of GPCR with an extracellular N-terminus, seven-transmembrane spanning regions and intracellular C-terminus. Sequence homology showed about 88% similarity to the human RDC1 orphan receptor. They also reported similarity between RDC1 and rabbit IL-8 receptor 2 (human CXCR2), whose amino acid sequence shares 32% identity and 43.3% similarity. While in mouse RDC1 gene resides on chromosome 1, in humans, ACKR3 is localized to Chromosome 2, where the genes coding for GPCRs CXCR1, CXCR2, and the co-receptor CXCR4 are localized (Ahuja, Ozcelik, Milatovitch, Francke, & Murphy, 1992; Libert et al., 1991; Moepps, Nuesseler, Braun, & Gierschik, 2006).

Two ligands, CXCL11 and CXCL12 bind and activate ACKR3. Burns et al. (2006) demonstrated the high-affinity binding of CXCL12 to human breast cancer cells (MCF7) using iodine-125 tracer labeled CXCL12 (^{125}I-SDF1α). However, CXCL11 also bound to cell surface ACKR3 at 10× lower affinity (Balabanian et al., 2005; Burns et al., 2006). It should be noted that all binding studies have used intact cells, including those with monoclonal antibodies to RDC1 (Balabanian et al., 2005). It was shown that both CXCL11 and CXCL12 bind to the same site on ACKR3 and therefore, CXCL11 could compete out CXCL12 (Burns et al., 2006). There are specific small molecule inhibitors of ACKR3 [e.g., CCX733 (Chemocentryx Inc., Mountain View, CA)], which block binding of CXCL12 to ACKR3, however do not interfere with CXCL12 binding to CXCR4, indicating unique binding site for CXCL12 on ACKR3 (Bachelerie, Ben-Baruch, et al., 2014; Burns et al., 2006; Gravel et al., 2010; Kalatskaya et al., 2009; Liang et al., 2019; Oishi et al., 2015; Xu, Li, Wu, Jiang, & Zhong, 2016). Binding of CXCL12 leads to homo- and heterodimerization of CXCR4 and ACKR3 when both receptors are co-expressed in cells, such as activated endothelial cells and some breast cancer cells (Luker, Gupta, & Luker, 2009; Luker, Gupta, Steele, Foerster, & Luker, 2009).

An elegant study by Luker et al., using luminescent ligands (luciferase-CXCL12), showed that upon ligand binding, ACKR3 dimerizes with CXCR4 (heterodimerization) and is rapidly internalized in a time and ligand dose-dependent manner (Luker, Gupta, & Luker, 2009). When cells are forced to express both receptors (ACKR3 and CXCR4) ligand stimulation leads to homodimerization (ACKR3-ACKR3 or CXCR4-CXCR4) and heterodimerization (ACKR3-CXCR4) (Luker, Gupta, Steele, et al., 2009). The authors reported rapid homodimerization but somewhat slow heterodimerization between ACKR3 and CXCR4 when treated with CXCL12. The slow heterodimerization is likely due to distinct changes in the conformation of the two receptors.

ACKR3 does not signal through G-proteins but is rather implicated in β-arrestin biased signaling. Our research demonstrated the opposing roles of β-arrestin 2 (ARRB2) in prostate and breast cancers. ARRB2 was a key regulator of ACKR3-EGFR interaction in breast cancer. Conversely, Src mediates ACKR3-induced transactivation of EGFR in prostate cancer cells, and this axis is negatively regulated by ARRB2 (Kallifatidis et al., 2016).

ACKR3 is also implicated in trans-endothelial migration (TEM) of tumor cells, which is a critical step in metastatic dissemination (Zabel, Lewen, Berahovich, Jaen, & Schall, 2011). Specifically, ACKR3 regulates TEM of CXCR4+/ACKR3+ tumor cells toward CXCL12. Furthermore, CXCL12 potentiates TEM toward chemokines expressed in endothelial cells within lymph nodes (such as CCL19 and CXCL13) in an ACKR3 dependent fashion (Zabel et al., 2011). ACKR3 is not present in the normal endothelium; however, it is upregulated in inflamed endothelial cells as well as endothelial cell-associated with tumor vasculature (Bachelerie, Ben-Baruch, et al., 2014; Burns et al., 2006; Liu, Carson-Walter, & Walter, 2014; Salazar et al., 2018; Yamada et al., 2015). Moreover, an acidic and hypoxic tumor microenvironment positively regulates ACKR3 expression in tumor-associated endothelial cells (TECs) (Salazar & Zabel, 2019), which regulates angiogenic properties through activation of ERK1/2 signaling (Yamada et al., 2015). In conclusion, ACKR3 plays a key role in tumor angiogenesis and metastasis (Fig. 4).

Several reports have demonstrated significant post-translational modifications of ACKR3 which are responsible for its varied functions. For example, ACKR3 is ubiquitinated in its membrane-bound status but is reversibly-deubiquitinated following binding to its ligand CXCL12. This process leads to internalization, degradation of the bound ligand in the acidic compartment of the cell, and recycling of ACKR3 to the cell surface, a process known as GPCR cycling (Canals et al., 2012). Hoffmann et al. reported that carboxyl-terminal serine/threonine residues are critical players in the scavenging of CXCL12 (Hoffmann et al., 2012). ACKR3 rapidly and avidly binds to CXCL12 in the extracellular matrix (due to 10× higher affinity compared to CXCR4), internalizes and degrades CXCL12. Canals et al. (2012) also report that the C-terminal lysine residue is critical for the reappearance of ACKR3 at the cell membrane.

2.3.4 ACKR3 and mitogenic signaling

The role of ACKR3 in contributing to more aggressive cancer phenotypes and therapeutic resistance may rely on alternative signaling mediated through

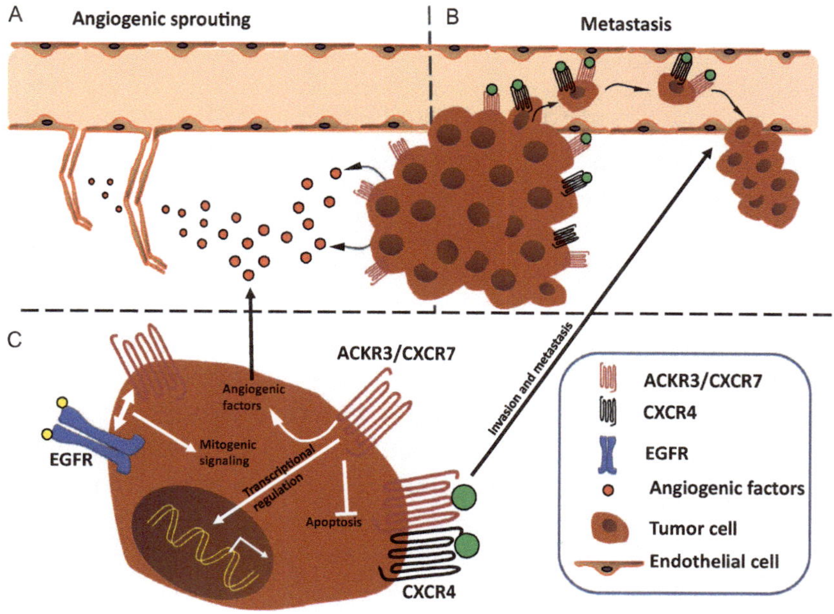

Fig. 4 Role of ACKR3 in cancer. CXCR7 is overexpressed in cancer cells and the tumor-associated endothelium. When expressed in tumor cells, it interacts with EGFR and mediates downstream mitogenic signaling. When expressed on endothelial cells in the tumor-associated vasculature it can be activated by CXCL12 released by the micro-environment and promote ERK1/2 signaling and angiogenesis (Yamada et al., 2015).

β-arrestin recruitment (Kallifatidis et al., 2016; Rajagopal et al., 2010). Published data have revealed that epidermal growth factor (EGF)-mediated activation of the EGF receptor (EGFR) is significantly enhanced with CXCR7 overexpression (Kallifatidis et al., 2016; Singh & Lokeshwar, 2011). Furthermore, this CXCR7-EGFR signaling axis is supported independently of ligand binding to CXCR7. We have shown that ACKR3 plays an essential role in prostate cancer cell proliferation, survival and clonogenic potential (Singh & Lokeshwar, 2011). Importantly, the androgen receptor (AR) interacts with the ACKR3 promoter region, suppressing its expression. However, androgen deprivation or AR inhibition induces ACKR3 expression and EGFR-mediated mitogenic signaling in prostate cancer (Hoy, Kallifatidis, Smith, & Lokeshwar, 2017). CRISPR/Cas9 mediated ACKR3 depletion abrogated EGFR-mediated signaling and induced cellular senescence (Hoy et al., 2017). Recently, Yu, Pham, Yu, and Wang (2019) reported that in addition to DHT, the Toll-like receptor 5 (TLR5) ligand flagellin, an inducer of prostatic inflammation, may also regulate ACKR3.

They report that flagellin decreased CXCR7 mRNA levels in the AR-dependent cell line LNCaP. The authors also report that while DHT and flagellin decrease ACKR3 protein, they additively increase the chemotactic migration of LNCaP cells toward CXCL12 (Yu et al., 2019). This report potentially conflicts the earlier consensus stating that ACKR3 is a scavenger receptor for CXCL12 thereby enhancing migration toward a CXCL12 source (Luker et al., 2012).

We reported recently that ACKR3 colocalizes with EGFR in breast cancer and drives EGF/EGFR-mediated mitogenic signaling (Salazar et al., 2014). By applying antibody proximity ligation assay (PLA), Salazar et al. reported that ACKR3 and EGFR form membrane aggregates when stimulated with EGF in situ. Interestingly, this clustering occurs in the absence of CXCL11 or CXCL12, indicating that it is mediated by conformational changes following the binding of EGF to EGFR. Further, the substitution of tyrosine_{1110} with phenylalanine abrogates this heterodimerization in the absence of CXCL12 (Kallifatidis et al., 2016). Overexpression of CXCR7 resulted in increased EGFR phosphorylation, specifically at tyrosine-1110 (Y_{1110}). Furthermore, it enhanced phosphorylation of ERK 1 and 2 (ERK1/2), leading to tumor cell proliferation and survival (Salazar et al., 2014; Singh & Lokeshwar, 2011).

While these investigations revealed that ACKR3 interacts with EGFR, the transactivation of EGFR required differential expression of β-arrestin-1 (ARRB1) and β-arrestin-2 (ARRB2). ARRB2 inhibited the interaction of ACKR3 with EGFR in prostate cancer. In that study, it was also demonstrated that the ACKR3-mediated EGFR transactivation was facilitated by *Src-Kinase* recruitment. This signaling axis, as well as downstream mitogenic signaling and cell proliferation, was enhanced following depletion of ARRB2, while overexpression of ARRB2 yielded an opposing effect (Kallifatidis et al., 2016). Other studies of β-arrestins and ACKR3 suggest an opposing role of ARRB1. Activation of EGFR in colorectal cancers leading to increased mitogenic signaling is mediated by ARRB1 recruitment to ACKR3 (Buchanan et al., 2006). Furthermore, ARRB1 facilitates ACKR3-mediated ERK signaling in glioma (in a CXCL12-dependent manner) (Odemis et al., 2012) as well as migration and mitogenic signaling mediated by CXCR4-ACKR3 heterodimers (Decaillot et al., 2011). Together these reports reveal an important link between increased ACKR3 expression and heightened mitogenic signaling and aggressive cancer phenotypes.

2.3.5 ACKR3 as a regulator of tumor cell migration

One of the principal functions of ACKR3 in the tumor microenvironment is its scavenging function. This function requires expression of CXCR4 in cancer cells and CXCL12 secretion from tumor stroma cells. It was shown that breast cancer cells that express a high level of CXCR4 metastasize to lung alveoli due to the high concentration of CXCL12 in lungs (Muller et al., 2001). A gradient of the chemoattractant concentration mediates the chemotaxis of CXCR4 expressing tumor cells to a tissue that secretes CXCL12. This gradient can be enhanced both spatially and temporally by sequestering, internalization/degradation of the chemoattractant by ACKR3. The "steepening" the CXCL12 gradient might increase the accumulation of tumor cells close to the CXCL12 source, in this case, the premetastatic site (Luker et al., 2012; Stacer et al., 2016). Wani et al. (2014) showed that CXCL12 enhances ACKR3 mediated breast cancer cell migration which resulted in increased metastasis. Their study further showed increased expression of ACKR3 in invasive ductal carcinoma and suggested ACKR3 as a prognostic marker for metastasis. Enhancement of metastatic potential by overexpression of ACKR3 in tumors is also associated with modification of tumor microenvironment by co-expression of pro-inflammatory chemokines. For example, high expression of ACKR3 alone is sufficient to induce a pro-inflammatory environment by increased expression of gelatinases (MMP-2 and MMP-9), and cell adhesion molecule (Wani et al., 2014). The interaction of tumor cells with the pro-inflammatory microenvironment might increase the expression of ACKR3 through a feedback mechanism. It was reported that the ACKR3 promoter has multiple binding sites of nuclear factor κB (NF-κB) and AP-1 both of which promote inflammation-mediated transcription of ACKR3 (Tarnowski et al., 2010). Singh and Lokeshwar reported that the pro-inflammatory chemokine CXCL8 triggers overexpression of ACKR3 in prostate cancer cells, leading to increased cell proliferation (Singh & Lokeshwar, 2011). Consistently, Kallifatidis et al. demonstrated increased cell motility following inhibition of β-arrestin 2 in LNCaP cells (Kallifatidis et al., 2016).

2.3.6 ACKR3 as a biomarker for tumor progression

ACKR3 is positively associated with prostate cancer progression. Wang et al. first reported that ACKR3 is increasingly expressed in castration-resistant and metastatic prostate cancer tissues (Wang, Chen, Guo, & Tao, 2018; Wang et al., 2008). Luo et al. reported that a combination of enzalutamide and ACKR3 inhibitor CCX771 inhibits castration-resistant AR-V7-expressing prostate cancer cells, therefore, indicating the existence of both androgen receptor-mediated and ACKR3 mediated mitogenic mechanisms in highly advanced prostate cancers (Luo et al., 2018). Further, it was reported that ACKR3 is responsible for the growth of castration-resistant prostate cancer by ACKR3 mediated activation of MAPK signaling pathway (Li et al., 2019). ACKR3 induced MAPK activation in other tumors such as lung and pancreatic cancer has also been reported (Becker et al., 2019; Liu et al., 2019). In KRAS-mutant non-small cell lung cancers (NSCLC), ACKR3 contributes to tumor growth in patients and offers a poor prognosis (Liu, Song, et al., 2019). In those cells, ACKR3 reactivates ERK signaling to promote ERK signaling and resistance to commonly used EGFR kinase domain inhibitors, such a gefitinib (Becker et al., 2019). These findings suggest ACKR3 expression and its versatile activities go beyond its function as a chemokine scavenger.

2.3.7 Inhibitors of ACKR3 activities

ACKR3 ligand-induced activity is inhibited by using either small molecule inhibitors such as the CCX773, which has shown efficacy in breast and bladder cancer models (Hao et al., 2012; Miao et al., 2007), and the llama derived single variable domain-directed antibodies in head and neck cancer models (Maussang et al., 2013). Glioblastoma, one of the most aggressive tumors, express high levels of ACKR3 and is negatively associated with patient survival (Salazar et al., 2018). A combination of a chimeric antibody against ACKR3 and temozolomide prolonged the life span of mice bearing an aggressive mouse glioblastoma (GL261) model (Salazar et al., 2018).

2.4 ACKR4/CCRL1

ACKR4 is present in keratinocytes, bronchial cells and epithelial cells and functions as a homeostatic chemokine scavenger (Borroni et al., 2018; Comerford et al., 2010; Gosling et al., 2000; Townson & Nibbs, 2002). The role of ACKR4 in tumor progression is controversial. Depletion/downregulation of ACKR4 resulted in worse disease outcomes in human hepatocellular, colon, and breast cancers, suggesting a protective role of ACKR4 in these tumor entities. However, there is also evidence that ACKR4 mediates EMT in breast cancer cells under certain circumstances and promotes lung metastasis. Further research needs to clarify these inconsistencies.

2.5 ACKR5/CCRL2

Only a few studies have investigated the role of ACKR5 in tumor progression. ACKR5 is present in the majority of hematopoietic cells such as macrophages, T cells, NK cells and monocytes (Galligan et al., 2004; Migeotte, Franssen, Goriely, Willems, & Parmentier, 2002; Patel, Charlton, Chambers, & Macphee, 2001). It may also be present in astrocytes, microglia and lung epithelial cells (H50–52). ACKR5 mediates the migration of lung dendritic cells toward lymph nodes (Yoshimura & Oppenheim, 2011). Importantly, ACKR5 can be constitutively expressed on endothelial cells; however, in some instances, pro-inflammatory stimuli induce its expression via NF-κB and JAK/STAT signaling. It functioned as a chemokine presenter (Borroni et al., 2018) and was found to bind the non-chemokine chemoattractant chemerin. When ACKR5 is expressed on endothelial cells, it binds to the chemerin without internalizing it, resulting in a local increase of the chemerin levels. This local increase of chemerin concentration mediates adhesion of lymphoid cells expressing CMKLR1—a second receptor for chemokine to sites of inflammation. Thus, ACKR5 may play a critical role in the regulation of trafficking of CMKLR1 expressing leucocytes toward chemokine presenting endothelia of inflammatory diseases.

Recent studies demonstrated the expression of ACKR5 in several cancers, including the tumors of colon and breast, as well as glioblastoma cell lines and tissues (Akram, Georges, Hielscher, Adwan, & Berger, 2016; Wang et al., 2015; Yin et al., 2012). It is also upregulated on prostate cancer cells compared to non-cancer cells (Reyes, Benedetti, Rebollo, Correa, & Geliebter, 2019). Similarly, ACKR5 is upregulated in cancerous acini compared to adjacent normal tissue (Reyes et al., 2019). The role of this receptor in cancer is controversial. Its expression is downregulated in highly invasive compared to minimally invasive breast cancer cell lines. Importantly, forced expression of ACKR5 invasive cell lines inhibited chemotaxis and invasion (Wang et al., 2015). Mechanistically, ACKR5 suppresses invasion in breast cancer through inhibition of CCL2-induced p38 phosphorylation (Wang et al., 2015), which is important for the downregulation of E-cadherin (Wang et al., 2015; Zohn et al., 2006). Conversely, ACKR5 is upregulated in human glioma samples in a grade-dependent manner with grade IV glioblastoma exhibiting the highest level of expression. Furthermore, ACKR5 overexpression induced migratory and invasion potential in glioblastoma cells (Yin et al., 2012; Yoshimura & Oppenheim, 2011).

2.6 ACKR6/PITPNM3/Nir1

ACKR6 is a conserved protein of the phosphatidylinositol transfer protein (PITP) family (Mobini et al., 2016). It is expressed in premature cells of the retina. ACKR6 is also expressed in other tissues, such as brain, spleen and ovaries (Krohn, Garin, Gabay, & Proudfoot, 2013; Lev, 2001; Ocaka, Spalluto, Wilson, Hunt, & Halford, 2005). When expressed on cancer cells it functions as an oncogene. Specifically, ACKR6 is a receptor for the chemokine CCL18 which is abundantly expressed and released by tumor-associated macrophages (Lin et al., 2015; Liu et al., 2019). ACKR6 is also expressed in endothelial cells and binding of CCL18 to its receptor promotes angiogenesis and tumor progression. Consistently, depletion of ACKR6 abrogated endothelial cell sprouting indicating the key role of ACKR6 in the angiogenic effects mediated by CCL18 (Lin et al., 2015). In breast cancer, expression of the ACKR6 ligand CCL18 positively correlates with micro-vessel density/tumor angiogenesis (Lin et al., 2015). Mechanistically, the CCL18/ACKR6 axis promotes angiogenesis through activation of ERK and Akt/GSK-3b/Snail signaling in endothelial cells (Lin et al., 2015). Li et al. showed that ACKR6 activation by its ligand induces calcium signaling and promotes migration of breast cancer cells (Li et al., 2014). CCL18 mediated activation of ACKR6 resulted in the phosphorylation and recruitment of proline-rich tyrosine kinase 2 (Pyk2)—a non-receptor cytoplasmic tyrosine kinase—to the cell membrane where it forms a complex with ACKR6. The CCL18 mediated ACKR6-Pyk2 complex formation induces activation of Src and plays a key role in regulating integrin alpha5/beta1 clustering and associated processes such as breast cancer cell adhesion, migration and invasion (Du et al., 2001; Li et al., 2014). Moreover, CCL18 binding to ACKR6 induces phosphorylation of LIMK and cofilin which induce actin polymerization and cytoskeleton rearrangement in breast cancer cells, processes that are essential for migration and invasion. Furthermore, activation of ACKR6 by its ligand induced the PI3K/Akt pathway and induces Snail mediated epithelial to mesenchymal transition (EMT) (Zhang et al., 2013).

Consistently, Zhang et al. showed that ACKR6 is expressed in the vast majority of invasive ductal carcinoma samples (91% ACKR6 expression) in contrast to normal mammary glands (6% ACKR6 expression) and is positively correlated with lymph node and distant metastasis (Zhang et al., 2013).

ACKR6 is also expressed in hepatocellular carcinoma (HCC) samples in a tumor grade-dependent fashion with a higher expression in grade III vs. grade I tumors. Similarly, ACKR6 is significantly overexpressed in HCC with metastasis compared to non-metastatic cases. Mechanistically, activation of ACKR6 by its ligand-induced NF-κB signaling in HCC cells and expression of NF-κB target gene Snail (Lin et al., 2016). The ACKR6 ligand CCL18 is also present in epithelial cells and tumor-associated macrophages of human pancreatic ductal adenocarcinoma (PDAC) where its expression correlates with lymph node metastasis, histopathological grading and overall survival (Meng et al., 2015). Interestingly, macrophage-secreted CCL18 induced the expression of the adhesion protein VCAM-1 in PDAC cells (Ye et al., 2018). VCAM-1 may promote a glycolytic phenotype in cancer cells since VCAM-1 depletion resulted in decreased glucose-uptake, glycolysis and lactate production (Ye et al., 2018). This macrophage mediated glycolytic phenotype in cancer cells results in an increased lactate secretion which serves as a paracrine signal and a positive feedback loop that regulates M2-like polarization of tumor-associated macrophages (Ye et al., 2018).

In conclusion, ACKR6 expression on cancer cells is associated with increased metastasis and worse prognosis. The receptor is activated by cytokines release from tumor-associated macrophages and induces downstream signaling that promotes EMT.

Acknowledgment

Authors' research is partially funded by the U.S. Veterans Health Administration Merit Award (5I01 BX003862-02).

References

Addison, C. L., Belperio, J. A., Burdick, M. D., & Strieter, R. M. (2004). Overexpression of the duffy antigen receptor for chemokines (DARC) by NSCLC tumor cells results in increased tumor necrosis. *BMC Cancer*, *4*, 28.

Ahuja, S. K., Ozcelik, T., Milatovitch, A., Francke, U., & Murphy, P. M. (1992). Molecular evolution of the human interleukin-8 receptor gene cluster. *Nature Genetics*, *2*, 31–36.

Akram, I. G., Georges, R., Hielscher, T., Adwan, H., & Berger, M. R. (2016). The chemokines CCR1 and CCRL2 have a role in colorectal cancer liver metastasis. *Tumour Biology*, *37*, 2461–2471.

Atzler, D., & de Winther, M. P. J. (2018). DARC matter(s) for inflammatory cells. *Cardiovascular Research*, *114*, e11–e13.

Bachelerie, F., Ben-Baruch, A., Burkhardt, A. M., Combadiere, C., Farber, J. M., Graham, G. J., et al. (2014). International Union of Basic and Clinical Pharmacology. [corrected]. LXXXIX. Update on the extended family of chemokine receptors and introducing a new nomenclature for atypical chemokine receptors. *Pharmacological Reviews*, *66*, 1–79.

Bachelerie, F., Graham, G. J., Locati, M., Mantovani, A., Murphy, P. M., Nibbs, R., et al. (2014). New nomenclature for atypical chemokine receptors. *Nature Immunology*, *15*, 207–208.

Balabanian, K., Lagane, B., Infantino, S., Chow, K. Y., Harriague, J., Moepps, B., et al. (2005). The chemokine SDF-1/CXCL12 binds to and signals through the orphan receptor RDC1 in T lymphocytes. *The Journal of Biological Chemistry*, *280*, 35760–35766.
Bandyopadhyay, S., Zhan, R., Chaudhuri, A., Watabe, M., Pai, S. K., Hirota, S., et al. (2006). Interaction of KAI1 on tumor cells with DARC on vascular endothelium leads to metastasis suppression. *Nature Medicine*, *12*, 933–938.
Banisadr, G., Podojil, J. R., Miller, S. D., & Miller, R. J. (2016). Pattern of CXCR7 gene expression in mouse brain under normal and inflammatory conditions. *Journal of Neuroimmune Pharmacology*, *11*, 26–35.
Becker, J. H., Gao, Y., Soucheray, M., Pulido, I., Kikuchi, E., Rodriguez, M. L., et al. (2019). CXCR7 reactivates ERK signaling to promote resistance to EGFR kinase inhibitors in NSCLC. *Cancer Research*, *79*, 4439–4452.
Berahovich, R. D., Zabel, B. A., Penfold, M. E., Lewen, S., Wang, Y., Miao, Z., et al. (2010). CXCR7 protein is not expressed on human or mouse leukocytes. *Journal of Immunology*, *185*, 5130–5139.
Bonini, J. A., Anderson, S. M., & Steiner, D. F. (1997). Molecular cloning and tissue expression of a novel orphan G protein-coupled receptor from rat lung. *Biochemical and Biophysical Research Communications*, *234*, 190–193.
Bonini, J. A., & Steiner, D. F. (1997). Molecular cloning and expression of a novel rat CC-chemokine receptor (rCCR10rR) that binds MCP-1 and MIP-1beta with high affinity. *DNA and Cell Biology*, *16*, 1023–1030.
Borroni, E. M., Savino, B., Bonecchi, R., & Locati, M. (2018). Chemokines sound the alarmin: The role of atypical chemokine in inflammation and cancer. *Seminars in Immunology*, *38*, 63–71.
Buchanan, F. G., Gorden, D. L., Matta, P., Shi, Q., Matrisian, L. M., & DuBois, R. N. (2006). Role of beta-arrestin 1 in the metastatic progression of colorectal cancer. *Proceedings of the National Academy of Sciences of the United States of America*, *103*, 1492–1497.
Burns, J. M., Summers, B. C., Wang, Y., Melikian, A., Berahovich, R., Miao, Z., et al. (2006). A novel chemokine receptor for SDF-1 and I-TAC involved in cell survival, cell adhesion, and tumor development. *The Journal of Experimental Medicine*, *203*, 2201–2213.
Canals, M., Scholten, D. J., de Munnik, S., Han, M. K., Smit, M. J., & Leurs, R. (2012). Ubiquitination of CXCR7 controls receptor trafficking. *PLoS One*, *7*, e34192.
Chaudhuri, A., Zbrzezna, V., Polyakova, J., Pogo, A. O., Hesselgesser, J., & Horuk, R. (1994). Expression of the Duffy antigen in K562 cells. Evidence that it is the human erythrocyte chemokine receptor. *The Journal of Biological Chemistry*, *269*, 7835–7838.
Comerford, I., Nibbs, R. J., Litchfield, W., Bunting, M., Harata-Lee, Y., Haylock-Jacobs, S., et al. (2010). The atypical chemokine receptor CCX-CKR scavenges homeostatic chemokines in circulation and tissues and suppresses Th17 responses. *Blood*, *116*, 4130–4140.
D'Alterio, C., Consales, C., Polimeno, M., Franco, R., Cindolo, L., Portella, L., et al. (2010). Concomitant CXCR4 and CXCR7 expression predicts poor prognosis in renal cancer. *Current Cancer Drug Targets*, *10*, 772–781.
Decaillot, F. M., Kazmi, M. A., Lin, Y., Ray-Saha, S., Sakmar, T. P., & Sachdev, P. (2011). CXCR7/CXCR4 heterodimer constitutively recruits beta-arrestin to enhance cell migration. *The Journal of Biological Chemistry*, *286*, 32188–32197.
Du, Q. S., Ren, X. R., Xie, Y., Wang, Q., Mei, L., & Xiong, W. C. (2001). Inhibition of PYK2-induced actin cytoskeleton reorganization, PYK2 autophosphorylation and focal adhesion targeting by FAK. *Journal of Cell Science*, *114*, 2977–2987.
Elson, J. K., Beebe-Dimmer, J. L., Morgenstern, H., Chilkuri, M., Blanchard, J., & Lentsch, A. B. (2011). The Duffy antigen/receptor for chemokines (DARC) and prostate-cancer risk among Jamaican men. *Journal of Immigrant and Minority Health*, *13*, 36–41.

Fra, A. M., Locati, M., Otero, K., Sironi, M., Signorelli, P., Massardi, M. L., et al. (2003). Cutting edge: Scavenging of inflammatory CC chemokines by the promiscuous putatively silent chemokine receptor D6. *Journal of Immunology*, *170*, 2279–2282.

Freitas, C., Desnoyer, A., Meuris, F., Bachelerie, F., Balabanian, K., & Machelon, V. (2014). The relevance of the chemokine receptor ACKR3/CXCR7 on CXCL12-mediated effects in cancers with a focus on virus-related cancers. *Cytokine & Growth Factor Reviews*, *25*, 307–316.

Gahan, J. C., Gosalbez, M., Yates, T., Young, E. E., Escudero, D. O., Chi, A., et al. (2012). Chemokine and chemokine receptor expression in kidney tumors: Molecular profiling of histological subtypes and association with metastasis. *The Journal of Urology*, *187*, 827–833.

Galligan, C. L., Matsuyama, W., Matsukawa, A., Mizuta, H., Hodge, D. R., Howard, O. M., et al. (2004). Up-regulated expression and activation of the orphan chemokine receptor, CCRL2, in rheumatoid arthritis. *Arthritis and Rheumatism*, *50*, 1806–1814.

Goldmann, T., Dromann, D., Radtke, J., Marwitz, S., Lang, D. S., Schultz, H., et al. (2008). CXCR7 transcription in human non-small cell lung cancer and tumor-free lung tissues; possible regulation upon chemotherapy. *Virchows Archiv*, *452*, 347–348.

Gosalbez, M., Hupe, M. C., Lokeshwar, S. D., Yates, T. J., Shields, J., Veerapen, M. K., et al. (2014). Differential expression of SDF-1 isoforms in bladder cancer. *The Journal of Urology*, *191*, 1899–1905.

Gosling, J., Dairaghi, D. J., Wang, Y., Hanley, M., Talbot, D., Miao, Z., et al. (2000). Cutting edge: Identification of a novel chemokine receptor that binds dendritic cell- and T cell-active chemokines including ELC, SLC, and TECK. *Journal of Immunology*, *164*, 2851–2856.

Graham, G. J., & Locati, M. (2013). Regulation of the immune and inflammatory responses by the 'atypical' chemokine receptor D6. *The Journal of Pathology*, *229*, 168–175.

Gravel, S., Malouf, C., Boulais, P. E., Berchiche, Y. A., Oishi, S., Fujii, N., et al. (2010). The peptidomimetic CXCR4 antagonist TC14012 recruits beta-arrestin to CXCR7: Roles of receptor domains. *The Journal of Biological Chemistry*, *285*, 37939–37943.

Grymula, K., Tarnowski, M., Wysoczynski, M., Drukala, J., Barr, F. G., Ratajczak, J., et al. (2010). Overlapping and distinct role of CXCR7-SDF-1/ITAC and CXCR4-SDF-1 axes in regulating metastatic behavior of human rhabdomyosarcomas. *International Journal of Cancer*, *127*, 2554–2568.

Han, J., Yoo, I., Lee, S., Jung, W., Kim, H. J., Hyun, S. H., et al. (2019). Atypical chemokine receptors 1, 2, 3 and 4: Expression and regulation in the endometrium during the estrous cycle and pregnancy and with somatic cell nucleus transfer-cloned embryos in pigs. *Theriogenology*, *129*, 121–129.

Hao, M., Zheng, J., Hou, K., Wang, J., Chen, X., Lu, X., et al. (2012). Role of chemokine receptor CXCR7 in bladder cancer progression. *Biochemical Pharmacology*, *84*, 204–214.

Hattermann, K., Held-Feindt, J., Lucius, R., Muerkoster, S. S., Penfold, M. E., Schall, T. J., et al. (2010). The chemokine receptor CXCR7 is highly expressed in human glioma cells and mediates antiapoptotic effects. *Cancer Research*, *70*, 3299–3308.

Heesen, M., Berman, M. A., Charest, A., Housman, D., Gerard, C., & Dorf, M. E. (1998). Cloning and chromosomal mapping of an orphan chemokine receptor: Mouse RDC1. *Immunogenetics*, *47*, 364–370.

Heinrich, E. L., Lee, W., Lu, J., Lowy, A. M., & Kim, J. (2012). Chemokine CXCL12 activates dual CXCR4 and CXCR7-mediated signaling pathways in pancreatic cancer cells. *Journal of Translational Medicine*, *10*, 68.

Hoffmann, F., Muller, W., Schutz, D., Penfold, M. E., Wong, Y. H., Schulz, S., et al. (2012). Rapid uptake and degradation of CXCL12 depend on CXCR7 carboxyl-terminal serine/threonine residues. *The Journal of Biological Chemistry*, *287*, 28362–28377.

Hong, R., Shen, M. H., Xie, X. H., & Ruan, S. M. (2012). Inhibition of breast cancer metastasis via PITPNM3 by pachymic acid. *Asian Pacific Journal of Cancer Prevention, 13,* 1877–1880.

Horton, L. W., Yu, Y., Zaja-Milatovic, S., Strieter, R. M., & Richmond, A. (2007). Opposing roles of murine duffy antigen receptor for chemokine and murine CXC chemokine receptor-2 receptors in murine melanoma tumor growth. *Cancer Research, 67,* 9791–9799.

Howes, R. E., Patil, A. P., Piel, F. B., Nyangiri, O. A., Kabaria, C. W., Gething, P. W., et al. (2011). The global distribution of the Duffy blood group. *Nature Communications, 2,* 266.

Hoy, J. J., Kallifatidis, G., Smith, D. K., & Lokeshwar, B. L. (2017). Inhibition of androgen receptor promotes CXC-chemokine receptor 7-mediated prostate cancer cell survival. *Scientific Reports,* 7, 3058.

Iiizumi, M., Bandyopadhyay, S., & Watabe, K. (2007). Interaction of Duffy antigen receptor for chemokines and KAI1: A critical step in metastasis suppression. *Cancer Research, 67,* 1411–1414.

Iwakiri, S., Mino, N., Takahashi, T., Sonobe, M., Nagai, S., Okubo, K., et al. (2009). Higher expression of chemokine receptor CXCR7 is linked to early and metastatic recurrence in pathological stage I nonsmall cell lung cancer. *Cancer, 115,* 2580–2593.

Kalatskaya, I., Berchiche, Y. A., Gravel, S., Limberg, B. J., Rosenbaum, J. S., & Heveker, N. (2009). AMD3100 is a CXCR7 ligand with allosteric agonist properties. *Molecular Pharmacology, 75,* 1240–1247.

Kallifatidis, G., Munoz, D., Singh, R. K., Salazar, N., Hoy, J. J., & Lokeshwar, B. L. (2016). Beta-arrestin-2 counters CXCR7-mediated EGFR transactivation and proliferation. *Molecular Cancer Research, 14,* 493–503.

Kapas, S., & Clark, A. J. (1995). Identification of an orphan receptor gene as a type 1 calcitonin gene-related peptide receptor. *Biochemical and Biophysical Research Communications, 217,* 832–838.

Klein, K. R., Karpinich, N. O., Espenschied, S. T., Willcockson, H. H., Dunworth, W. P., Hoopes, S. L., et al. (2014). Decoy receptor CXCR7 modulates adrenomedullin-mediated cardiac and lymphatic vascular development. *Developmental Cell, 30,* 528–540.

Kollmar, O., Rupertus, K., Scheuer, C., Nickels, R. M., Haberl, G. C., Tilton, B., et al. (2010). CXCR4 and CXCR7 regulate angiogenesis and CT26.WT tumor growth independent from SDF-1. *International Journal of Cancer, 126,* 1302–1315.

Krohn, S., Garin, A., Gabay, C., & Proudfoot, A. E. L. (2013). The activity of CCL18 is principally mediated through interaction with glycosaminoglycans. *Frontiers in Immunology, 4,* 1–15 (Article 193).

Langenes, V., Svensson, H., Borjesson, L., Gustavsson, B., Bemark, M., Sjoling, A., et al. (2013). Expression of the chemokine decoy receptor D6 is decreased in colon adenocarcinomas. *Cancer Immunology, Immunotherapy, 62,* 1687–1695.

Lee, H. J., Song, I. C., Yun, H. J., Jo, D. Y., & Kim, S. (2014). CXC chemokines and chemokine receptors in gastric cancer: From basic findings towards therapeutic targeting. *World Journal of Gastroenterology, 20,* 1681–1693.

Lev, S. (2001). Molecular aspects of retinal degenerative diseases. *Cellular and Molecular Neurobiology, 21,* 575–589.

Li, H. Y., Cui, X. Y., Wu, W., Yu, F. Y., Yao, H. R., Liu, Q., et al. (2014). Pyk2 and Src mediate signaling to CCL18-induced breast cancer metastasis. *Journal of Cellular Biochemistry, 115,* 596–603.

Li, S., Fong, K. W., Gritsina, G., Zhang, A., Zhao, J. C., Kim, J., et al. (2019). Activation of MAPK signaling by CXCR7 leads to enzalutamide resistance in prostate cancer. *Cancer Research, 79,* 2580–2592.

Liang, X., Luo, D., Yan, J. L., Rezaei, M. A., Salvador-Reyes, L. A., Gunasekera, S. P., et al. (2019). Discovery of amantamide, a selective CXCR7 agonist from marine cyanobacteria. *Organic Letters*, *21*, 1622–1626.
Liberman, J., Sartelet, H., Flahaut, M., Muhlethaler-Mottet, A., Coulon, A., Nyalendo, C., et al. (2012). Involvement of the CXCR7/CXCR4/CXCL12 axis in the malignant progression of human neuroblastoma. *PLoS One*, *7*, e43665.
Libert, F., Parmentier, M., Lefort, A., Dumont, J. E., & Vassart, G. (1990). Complete nucleotide sequence of a putative G protein coupled receptor: RDC1. *Nucleic Acids Research*, *18*, 1917.
Libert, F., Passage, E., Parmentier, M., Simons, M. J., Vassart, G., & Mattei, M. G. (1991). Chromosomal mapping of A1 and A2 adenosine receptors, VIP receptor, and a new subtype of serotonin receptor. *Genomics*, *11*, 225–227.
Lin, L., Chen, Y. S., Yao, Y. D., Chen, J. Q., Chen, J. N., Huang, S. Y., et al. (2015). CCL18 from tumor-associated macrophages promotes angiogenesis in breast cancer. *Oncotarget*, *6*, 34758–34773.
Lin, Z., Li, W., Zhang, H., Wu, W., Peng, Y., Zeng, Y., et al. (2016). CCL18/PITPNM3 enhances migration, invasion, and EMT through the NF-kappaB signaling pathway in hepatocellular carcinoma. *Tumour Biology*, *37*, 3461–3468.
Liu, Y., Carson-Walter, E., & Walter, K. A. (2014). Chemokine receptor CXCR7 is a functional receptor for CXCL12 in brain endothelial cells. *PLoS One*, *9*, e103938.
Liu, B., Song, S., Setroikromo, R., Chen, S., Hu, W., Chen, D., et al. (2019). CX chemokine receptor 7 contributes to survival of KRAS-mutant non-small cell lung cancer upon loss of epidermal growth factor receptor. *Cancers (Basel)*, *11*, 455.
Liu, Y., Zheng, H., Li, Q., Li, S., Lai, H., Song, E., et al. (2019). Discovery of CCL18 antagonist blocking breast cancer metastasis. *Clinical & Experimental Metastasis*, *36*, 243–255.
Luker, K., Gupta, M., & Luker, G. (2009). Bioluminescent CXCL12 fusion protein for cellular studies of CXCR4 and CXCR7. *BioTechniques*, *47*, 625–632.
Luker, K. E., Gupta, M., Steele, J. M., Foerster, B. R., & Luker, G. D. (2009). Imaging ligand-dependent activation of CXCR7. *Neoplasia*, *11*, 1022–1035.
Luker, K. E., Lewin, S. A., Mihalko, L. A., Schmidt, B. T., Winkler, J. S., Coggins, N. L., et al. (2012). Scavenging of CXCL12 by CXCR7 promotes tumor growth and metastasis of CXCR4-positive breast cancer cells. *Oncogene*, *31*, 4750–4758.
Luker, K. E., Steele, J. M., Mihalko, L. A., Ray, P., & Luker, G. D. (2010). Constitutive and chemokine-dependent internalization and recycling of CXCR7 in breast cancer cells to degrade chemokine ligands. *Oncogene*, *29*, 4599–4610.
Luo, Y., Azad, A. K., Karanika, S., Basourakos, S. P., Zuo, X., Wang, J., et al. (2018). Enzalutamide and CXCR7 inhibitor combination treatment suppresses cell growth and angiogenic signaling in castration-resistant prostate cancer models. *International Journal of Cancer*, *142*, 2163–2174.
Maeda, S., Kuboki, S., Nojima, H., Shimizu, H., Yoshitomi, H., Furukawa, K., et al. (2017). Duffy antigen receptor for chemokines (DARC) expressing in cancer cells inhibits tumor progression by suppressing CXCR2 signaling in human pancreatic ductal adenocarcinoma. *Cytokine*, *95*, 12–21.
Maksym, R. B., Tarnowski, M., Grymula, K., Tarnowska, J., Wysoczynski, M., Liu, R., et al. (2009). The role of stromal-derived factor-1—CXCR7 axis in development and cancer. *European Journal of Pharmacology*, *625*, 31–40.
Martinez de la Torre, Y., Buracchi, C., Borroni, E. M., Dupor, J., Bonecchi, R., Nebuloni, M., et al. (2007). Protection against inflammation- and autoantibody-caused fetal loss by the chemokine decoy receptor D6. *Proceedings of the National Academy of Sciences of the United States of America*, *104*, 2319–2324.
Massara, M., Bonavita, O., Mantovani, A., Locati, M., & Bonecchi, R. (2016). Atypical chemokine receptors in cancer: Friends or foes? *Journal of Leukocyte Biology*, *99*, 927–933.

Maussang, D., Mujic-Delic, A., Descamps, F. J., Stortelers, C., Vanlandschoot, P., Stigter-van Walsum, M., et al. (2013). Llama-derived single variable domains (nanobodies) directed against chemokine receptor CXCR7 reduce head and neck cancer cell growth in vivo. *The Journal of Biological Chemistry*, *288*, 29562–29572.
McKimmie, C. S., Fraser, A. R., Hansell, C., Gutierrez, L., Philipsen, S., Connell, L., et al. (2008). Hemopoietic cell expression of the chemokine decoy receptor D6 is dynamic and regulated by GATA1. *Journal of Immunology*, *181*, 8171–8181.
McKimmie, C. S., Singh, M. D., Hewit, K., Lopez-Franco, O., Le Brocq, M., Rose-John, S., et al. (2013). An analysis of the function and expression of D6 on lymphatic endothelial cells. *Blood*, *121*, 3768–3777.
Meijer, J., Ogink, J., & Roos, E. (2008). Effect of the chemokine receptor CXCR7 on proliferation of carcinoma cells in vitro and in vivo. *British Journal of Cancer*, *99*, 1493–1501.
Meng, F., Li, W., Li, C., Gao, Z., Guo, K., & Song, S. (2015). CCL18 promotes epithelial-mesenchymal transition, invasion and migration of pancreatic cancer cells in pancreatic ductal adenocarcinoma. *International Journal of Oncology*, *46*, 1109–1120.
Miao, Z., Luker, K. E., Summers, B. C., Berahovich, R., Bhojani, M. S., Rehemtulla, A., et al. (2007). CXCR7 (RDC1) promotes breast and lung tumor growth in vivo and is expressed on tumor-associated vasculature. *Proceedings of the National Academy of Sciences of the United States of America*, *104*, 15735–15740.
Migeotte, I., Franssen, J. D., Goriely, S., Willems, F., & Parmentier, M. (2002). Distribution and regulation of expression of the putative human chemokine receptor HCR in leukocyte populations. *European Journal of Immunology*, *32*, 494–501.
Mitchell, A., Hasanali, S. L., Morera, D. S., Baskar, R., Wang, X., Khan, R., et al. (2019). A chemokine/chemokine receptor signature potentially predicts clinical outcome in colorectal cancer patients. *Cancer Biomarkers*, *26*, 291–301.
Mobini, G. R., Ghahremani, M. H., Amanpour, S., Dehpour, A. R., Akbari, A., Hoseiniharouni, S. M., et al. (2016). Transforming growth factor beta-induced factor 2-linked X (TGIF2LX) regulates two morphogenesis genes, Nir1 and Nir2 in human colorectal. *Acta Medica Iranica*, *54*, 302–307.
Moepps, B., Nuesseler, E., Braun, M., & Gierschik, P. (2006). A homolog of the human chemokine receptor CXCR1 is expressed in the mouse. *Molecular Immunology*, *43*, 897–914.
Monnier, J., Boissan, M., L'Helgoualc'h, A., Lacombe, M. L., Turlin, B., Zucman-Rossi, J., et al. (2012). CXCR7 is up-regulated in human and murine hepatocellular carcinoma and is specifically expressed by endothelial cells. *European Journal of Cancer*, *48*, 138–148.
Muller, A., Homey, B., Soto, H., Ge, N., Catron, D., Buchanan, M. E., et al. (2001). Involvement of chemokine receptors in breast cancer metastasis. *Nature*, *410*, 50–56.
Nagata, S., Ishihara, T., Robberecht, P., Libert, F., Parmentier, M., Christophe, J., et al. (1992). RDC1 may not be VIP receptor. *Trends in Pharmacological Sciences*, *13*, 102–103.
Nemesure, B., Wu, S. Y., Hennis, A., & Leske, M. C. (2015). Distribution of Duffy antigen receptor for chemokines (DARC) and risk of prostate cancer in Barbados, West Indies. *Journal of Immigrant and Minority Health*, *17*, 679–683.
Neote, K., DiGregorio, D., Mak, J. Y., Horuk, R., & Schall, T. J. (1993). Molecular cloning, functional expression, and signaling characteristics of a C-C chemokine receptor. *Cell*, *72*, 415–425.
Neote, K., Mak, J. Y., Kolakowski, L. F., Jr., & Schall, T. J. (1994). Functional and biochemical analysis of the cloned Duffy antigen: Identity with the red blood cell chemokine receptor. *Blood*, *84*, 44–52.
Nibbs, R. J., Gilchrist, D. S., King, V., Ferra, A., Forrow, S., Hunter, K. D., et al. (2007). The atypical chemokine receptor D6 suppresses the development of chemically induced skin tumors. *The Journal of Clinical Investigation*, *117*, 1884–1892.
Nibbs, R. J., & Graham, G. J. (2013). Immune regulation by atypical chemokine receptors. *Nature Reviews. Immunology*, *13*, 815–829.

Nibbs, R. J., Wylie, S. M., Pragnell, I. B., & Graham, G. J. (1997). Cloning and characterization of a novel murine beta chemokine receptor, D6. Comparison to three other related macrophage inflammatory protein-1alpha receptors, CCR-1, CCR-3, and CCR-5. *The Journal of Biological Chemistry*, *272*, 12495–12504.

Ocaka, L., Spalluto, C., Wilson, D. I., Hunt, D. M., & Halford, S. (2005). Chromosomal localization, genomic organization and evolution of the genes encoding human phosphatidylinositol transfer protein membrane-associated (PITPNM) 1, 2 and 3. *Cytogenetic and Genome Research*, *108*, 293–302.

Odemis, V., Lipfert, J., Kraft, R., Hajek, P., Abraham, G., Hattermann, K., et al. (2012). The presumed atypical chemokine receptor CXCR7 signals through G(i/o) proteins in primary rodent astrocytes and human glioma cells. *Glia*, *60*, 372–381.

Oishi, S., Kuroyanagi, T., Kubo, T., Montpas, N., Yoshikawa, Y., Misu, R., et al. (2015). Development of novel CXC chemokine receptor 7 (CXCR7) ligands: Selectivity switch from CXCR4 antagonists with a cyclic pentapeptide scaffold. *Journal of Medicinal Chemistry*, *58*, 5218–5225.

Patel, L., Charlton, S. J., Chambers, J. K., & Macphee, C. H. (2001). Expression and functional analysis of chemokine receptors in human peripheral blood leukocyte populations. *Cytokine*, *14*, 27–36.

Puchert, M., Pelkner, F., Stein, G., Angelov, D. N., Boltze, J., Wagner, D. C., et al. (2017). Astrocytic expression of the CXCL12 receptor, CXCR7/ACKR3 is a hallmark of the diseased, but not developing CNS. *Molecular and Cellular Neurosciences*, *85*, 105–118.

Rajagopal, S., Kim, J., Ahn, S., Craig, S., Lam, C. M., Gerard, N. P., et al. (2010). Beta-arrestin- but not G protein-mediated signaling by the "decoy" receptor CXCR7. *Proceedings of the National Academy of Sciences of the United States of America*, *107*, 628–632.

Reyes, N., Benedetti, I., Rebollo, J., Correa, O., & Geliebter, J. (2019). Atypical chemokine receptor CCRL2 is overexpressed in prostate cancer cells. *The Journal of Biomedical Research*, *33*(1), 17–23.

Rot, A. (2005). Contribution of Duffy antigen to chemokine function. *Cytokine & Growth Factor Reviews*, *16*, 687–694.

Salazar, N., Carlson, J. C., Huang, K., Zheng, Y., Oderup, C., Gross, J., et al. (2018). A chimeric antibody against ACKR3/CXCR7 in combination with TMZ activates immune responses and extends survival in mouse GBM models. *Molecular Therapy*, *26*, 1354–1365.

Salazar, N., Munoz, D., Kallifatidis, G., Singh, R. K., Jorda, M., & Lokeshwar, B. L. (2014). The chemokine receptor CXCR7 interacts with EGFR to promote breast cancer cell proliferation. *Molecular Cancer*, *13*, 198.

Salazar, N., & Zabel, B. A. (2019). Support of tumor endothelial cells by chemokine receptors. *Frontiers in Immunology*, *10*, 147.

Salmaggi, A., Maderna, E., Calatozzolo, C., Gaviani, P., Canazza, A., Milanesi, I., et al. (2009). CXCL12, CXCR4 and CXCR7 expression in brain metastases. *Cancer Biology & Therapy*, *8*, 1608–1614.

Savino, B., Caronni, N., Anselmo, A., Pasqualini, F., Borroni, E. M., Basso, G., et al. (2014). ERK-dependent downregulation of the atypical chemokine receptor D6 drives tumor aggressiveness in Kaposi sarcoma. *Cancer Immunology Research*, *2*, 679–689.

Schneider, C., Teufel, A., Yevsa, T., Staib, F., Hohmeyer, A., Walenda, G., et al. (2012). Adaptive immunity suppresses formation and progression of diethylnitrosamine-induced liver cancer. *Gut*, *61*, 1733–1743.

Shen, H., Schuster, R., Stringer, K. F., Waltz, S. E., & Lentsch, A. B. (2006). The Duffy antigen/receptor for chemokines (DARC) regulates prostate tumor growth. *The FASEB Journal*, *20*, 59–64.

Sierro, F., Biben, C., Martinez-Munoz, L., Mellado, M., Ransohoff, R. M., Li, M., et al. (2007). Disrupted cardiac development but normal hematopoiesis in mice deficient in the second CXCL12/SDF-1 receptor, CXCR7. *Proceedings of the National Academy of Sciences of the United States of America, 104*, 14759–14764.

Singh, R. K., & Lokeshwar, B. L. (2011). The IL-8-regulated chemokine receptor CXCR7 stimulates EGFR signaling to promote prostate cancer growth. *Cancer Research, 71*, 3268–3277.

Sreedharan, S. P., Robichon, A., Peterson, K. E., & Goetzl, E. J. (1991). Cloning and expression of the human vasoactive intestinal peptide receptor. *Proceedings of the National Academy of Sciences of the United States of America, 88*, 4986–4990.

Stacer, A. C., Fenner, J., Cavnar, S. P., Xiao, A., Zhao, S., Chang, S. L., et al. (2016). Endothelial CXCR7 regulates breast cancer metastasis. *Oncogene, 35*, 1716–1724.

Su, A. I., Cooke, M. P., Ching, K. A., Hakak, Y., Walker, J. R., Wiltshire, T., et al. (2002). Large-scale analysis of the human and mouse transcriptomes. *Proceedings of the National Academy of Sciences of the United States of America, 99*, 4465–4470.

Sun, X., Cheng, G., Hao, M., Zheng, J., Zhou, X., Zhang, J., et al. (2010). CXCL12/CXCR4/CXCR7 chemokine axis and cancer progression. *Cancer Metastasis Reviews, 29*, 709–722.

Sun, Y. X., Wang, J., Shelburne, C. E., Lopatin, D. E., Chinnaiyan, A. M., Rubin, M. A., et al. (2003). Expression of CXCR4 and CXCL12 (SDF-1) in human prostate cancers (PCa) in vivo. *Journal of Cellular Biochemistry, 89*, 462–473.

Tarnowski, M., Grymula, K., Reca, R., Jankowski, K., Maksym, R., Tarnowska, J., et al. (2010). Regulation of expression of stromal-derived factor-1 receptors: CXCR4 and CXCR7 in human rhabdomyosarcomas. *Molecular Cancer Research, 8*, 1–14.

Townson, J. R., & Nibbs, R. J. (2002). Characterization of mouse CCX-CKR, a receptor for the lymphocyte-attracting chemokines TECK/mCCL25, SLC/mCCL21 and MIP-3beta/mCCL19: Comparison to human CCX-CKR. *European Journal of Immunology, 32*, 1230–1241.

Vetrano, S., Borroni, E. M., Sarukhan, A., Savino, B., Bonecchi, R., Correale, C., et al. (2010). The lymphatic system controls intestinal inflammation and inflammation-associated colon cancer through the chemokine decoy receptor D6. *Gut, 59*, 197–206.

Virgintino, D., Errede, M., Rizzi, M., Girolamo, F., Strippoli, M., Walchli, T., et al. (2013). The CXCL12/CXCR4/CXCR7 ligand-receptor system regulates neuro-glio-vascular interactions and vessel growth during human brain development. *Journal of Inherited Metabolic Disease, 36*, 455–466.

Wang, L. P., Cao, J., Zhang, J., Wang, B. Y., Hu, X. C., Shao, Z. M., et al. (2015). The human chemokine receptor CCRL2 suppresses chemotaxis and invasion by blocking CCL2-induced phosphorylation of p38 MAPK in human breast cancer cells. *Medical Oncology, 32*, 254.

Wang, W., Chen, Z. X., Guo, D. Y., & Tao, Y. X. (2018). Regulation of prostate cancer by hormone-responsive G protein-coupled receptors. *Pharmacology & Therapeutics, 191*, 135–147.

Wang, J., Ou, Z. L., Hou, Y. F., Luo, J. M., Shen, Z. Z., Ding, J., et al. (2006). Enhanced expression of Duffy antigen receptor for chemokines by breast cancer cells attenuates growth and metastasis potential. *Oncogene, 25*, 7201–7211.

Wang, J., Shiozawa, Y., Wang, J., Wang, Y., Jung, Y., Pienta, K. J., et al. (2008). The role of CXCR7/RDC1 as a chemokine receptor for CXCL12/SDF-1 in prostate cancer. *The Journal of Biological Chemistry, 283*, 4283–4294.

Wani, N., Nasser, M. W., Ahirwar, D. K., Zhao, H., Miao, Z., Shilo, K., et al. (2014). C-X-C motif chemokine 12/C-X-C chemokine receptor type 7 signaling regulates breast cancer growth and metastasis by modulating the tumor microenvironment. *Breast Cancer Research, 16*, R54.

Wu, F. Y., Ou, Z. L., Feng, L. Y., Luo, J. M., Wang, L. P., Shen, Z. Z., et al. (2008). Chemokine decoy receptor d6 plays a negative role in human breast cancer. *Molecular Cancer Research, 6*, 1276–1288.

Wurth, R., Barbieri, F., Bajetto, A., Pattarozzi, A., Gatti, M., Porcile, C., et al. (2011). Expression of CXCR7 chemokine receptor in human meningioma cells and in intratumoral microvasculature. *Journal of Neuroimmunology, 234*, 115–123.

Xu, D., Li, R., Wu, J., Jiang, L., & Zhong, H. A. (2016). Drug design targeting the CXCR4/CXCR7/CXCL12 pathway. *Current Topics in Medicinal Chemistry, 16*, 1441–1451.

Xue, T. C., Chen, R. X., Han, D., Chen, J., Xue, Q., Gao, D. M., et al. (2012). Downregulation of CXCR7 inhibits the growth and lung metastasis of human hepatocellular carcinoma cells with highly metastatic potential. *Experimental and Therapeutic Medicine, 3*, 117–123.

Yamada, K., Maishi, N., Akiyama, K., Towfik Alam, M., Ohga, N., Kawamoto, T., et al. (2015). CXCL12-CXCR7 axis is important for tumor endothelial cell angiogenic property. *International Journal of Cancer, 137*, 2825–2836.

Yang, C., Yu, K. D., Xu, W. H., Chen, A. X., Fan, L., Ou, Z. L., et al. (2013). Effect of genetic variants in two chemokine decoy receptor genes, DARC and CCBP2, on metastatic potential of breast cancer. *PLoS One, 8*, e78901.

Yates, T. J., Knapp, J., Gosalbez, M., Lokeshwar, S. D., Gomez, C. S., Benitez, A., et al. (2013). C-X-C chemokine receptor 7: A functionally associated molecular marker for bladder cancer. *Cancer, 119*, 61–71.

Ye, H., Zhou, Q., Zheng, S., Li, G., Lin, Q., Wei, L., et al. (2018). Tumor-associated macrophages promote progression and the Warburg effect via CCL18/NF-kB/VCAM-1 pathway in pancreatic ductal adenocarcinoma. *Cell Death & Disease, 9*, 453.

Yin, F., Xu, Z., Wang, Z., Yao, H., Shen, Z., Yu, F., et al. (2012). Elevated chemokine CC-motif receptor-like 2 (CCRL2) promotes cell migration and invasion in glioblastoma. *Biochemical and Biophysical Research Communications, 429*, 168–172.

Yoshimura, T., & Oppenheim, J. J. (2011). Chemokine-like receptor 1 (CMKLR1) and chemokine (C-C motif) receptor-like 2 (CCRL2); two multifunctional receptors with unusual properties. *Experimental Cell Research, 317*, 674–684.

Yu, L., Pham, Q., Yu, L. L., & Wang, T. T. Y. (2019). Modulation of CXC-motif chemokine receptor 7, but not 4, expression is related to migration of the human prostate cancer cell LNCaP: Regulation by androgen and inflammatory stimuli. *Inflammation Research* (PMID 31865399).

Yu, K. D., Wang, X., Yang, C., Zeng, X. H., & Shao, Z. M. (2015). Host genotype and tumor phenotype of chemokine decoy receptors integrally affect breast cancer relapse. *Oncotarget, 6*, 26519–26527.

Zabel, B. A., Lewen, S., Berahovich, R. D., Jaen, J. C., & Schall, T. J. (2011). The novel chemokine receptor CXCR7 regulates trans-endothelial migration of cancer cells. *Molecular Cancer, 10*, 73.

Zeng, X. H., Ou, Z. L., Yu, K. D., Feng, L. Y., Yin, W. J., Li, J., et al. (2011). Coexpression of atypical chemokine binders (ACBs) in breast cancer predicts better outcomes. *Breast Cancer Research and Treatment, 125*, 715–727.

Zeng, X. H., Ou, Z. L., Yu, K. D., Feng, L. Y., Yin, W. J., Li, J., et al. (2014). Absence of multiple atypical chemokine binders (ACBs) and the presence of VEGF and MMP-9 predict axillary lymph node metastasis in early breast carcinomas. *Medical Oncology, 31*, 145.

Zhang, B., Yin, C., Li, H., Shi, L., Liu, N., Sun, Y., et al. (2013). Nir1 promotes invasion of breast cancer cells by binding to chemokine (C-C motif) ligand 18 through the PI3K/Akt/GSK3beta/Snail signalling pathway. *European Journal of Cancer, 49*, 3900–3913.

Zheng, K., Li, H. Y., Su, X. L., Wang, X. Y., Tian, T., Li, F., et al. (2010). Chemokine receptor CXCR7 regulates the invasion, angiogenesis and tumor growth of human hepatocellular carcinoma cells. *Journal of Experimental & Clinical Cancer Research, 29*, 31.

Zohn, I. E., Li, Y., Skolnik, E. Y., Anderson, K. V., Han, J., & Niswander, L. (2006). p38 and a p38-interacting protein are critical for downregulation of E-cadherin during mouse gastrulation. *Cell, 125*, 957–969.

CHAPTER TWO

The G protein coupled receptor CCR5 in cancer

Chandan Upadhyaya[a,c], Xuanmao Jiao[a], Anthony Ashton[a,d], Kishan Patel[a,c], Andrew V. Kossenkov[b], Richard G. Pestell[a,b,c,*]

[a]Pennsylvania Cancer and Regenerative Medicine Research Center, Baruch S. Blumberg Institute, Pennsylvania Biotechnology Center, Wynnewood, PA, United States
[b]Wistar Institute, Philadelphia, PA, United States
[c]Xavier University School of Medicine, Woodbury, NY, United States
[d]Division of Perinatal Research, Royal North Shore Hospital, St. Leonards, NSW, Australia
*Corresponding author: e-mail address: richard.pestell@bblumberg.org

Contents

Abstract

The G coupled protein receptor C—C chemokine receptor type 5 (CCR5) has the unusual characteristic in humans of being a developmentally non-essential gene that participates in several pathological processes including infection with HIV (Dean et al., 1996; Gupta et al., 2019; Samson et al., 1996), progression of stroke (Joy et al., 2019), osteoporosis (Xie et al., 2019) and the metastasis of cancer (Jiao et al., 2018; Velasco-Velazquez et al. 2012, 2014) (Reviewed in: Jiao, Nawab, et al., 2019; Jiao, Wang, & Pestell, 2019). The importance of CCR5 in HIV led to recent genetic engineering of humans to recreate a non-functional CCR5 gene. Thus, although the application of gene-editing tools, to manipulate human embryos is prohibited in the United States, and China. at the Second International Summit on Human Genome Editing in Hong Kong (http://www.nationalacademies.org/), it was claimed that CRISPR-Cas9 systems

Advances in Cancer Research, Volume 145
ISSN 0065-230X
https://doi.org/10.1016/bs.acr.2019.11.001

had been used to edit the *CCR5* gene in twin baby girls. The importance of CCR5 in stroke has led to clinical trials using maraviroc (NCT03172026). The key function of CCR5 in cancer metastasis and homing (Jiao et al., 2018; Jiao, Nawab, et al., 2019; Velasco-Velazquez et al. 2012, 2014) has led to three active clinical trials for metastatic cancer using CCR5 antagonists (Jiao, Nawab, et al., 2019; Jiao, Wang, & Pestell, 2019). Thus, it was surprising to find that the all-cause mortality rate in individuals who are homozygous for the CCR5Δ32 allele in the United Kingdom normal population was increased >20% increase, with an almost 2 year reduction overall lifespan (Wei & Nielsen, 2019). The current review herein discusses the distinct functions of CCR5 in human disease and potential avenues for further research.

1. Genetics of cysteine-cysteine chemokine receptor type (CCR5)

Interestingly, a homozygous gene deletion within the CCR5 coding region occurs naturally in humans (CCR5Δ32). These individuals remain asymptomatic throughout most of their lives. When introduced to pathogens individuals with CCR5Δ32 have resistance to HIV (Dean et al., 1996; Samson et al., 1996) poxvirus (Lalani et al., 1999) and a *Staphylococcus Aureus* produced leukotoxin (LukED) (Alonzo et al., 2013). In contrast CCR5Δ32 individuals have reduced resistance to West Nile virus (Lim et al., 2008), and tickborne encephalitis (Kindberg et al., 2008). LukED binds to CCR5 on T cell macrophages and dendritic cells (Alonzo et al., 2013) and may have participated in selecting for the genetic variant, which is more prevalent in individuals of European (10%) compared with individuals of Asian, Middle Eastern (2-to 5%) or African origin. Studies of mortality amongst individuals with CCR5Δ32 showed the all-cause mortality rate in individuals who are homozygous for the CCR5Δ32 allele was increased >20% increase, with an almost 2 year reduction overall lifespan (Wei & Nielsen, 2019). In addition to the CCR5Δ32 polymorphism, other nonfunctional or hypofunctional CCR5 alleles are relatively frequent in human populations (Blanpain et al., 2000) and their importance in disease predisposition and overall mortality has not been determined.

2. CCR5 and biochemical signaling

The signal transduction elicited by CCR5 appears to be relatively well conserved across cell types in which physiological expression (inflammatory cells) or ectopic expression (cancer cells) has been assessed. CCR5 became a focus of much interest because of its role as a co-receptor for HIV entry. In response to HIV infection, CCR5 is deployed as a co-receptor. The HIV-1 envelope glycoprotein (Env), which consists of trimeric $(gp160)_3$ cleaved to $(gp120$ and $gp41)_3$, interacts with the primary receptor CD4 and the coreceptor (CCR5) to fuse viral and target-cell membranes. The structure of CCR5 bound to CD4, suggests CCR5 anchors the CD4-induced conformation of Env near the cell membrane (Shaik et al., 2019).

In response to ligands, however, GPCR undergoes a conformational change, forcing the Gαi and Gβγ subunits to dissociate. CCR5 is a 7 trans-membrane G-protein coupled receptor (GPCR) that binds to multiple ligands including (CCL3 (MIP1α), CXCL13 (BCA-1), CCL4 (MP-1β), CCL3L1, CCL8 (MCP2), CCL5 (RANTES), CCL11 (Eotaxin), CCL13 (MCP-4), and CCL16 (HCC-4) (Fig. 1) (Jiao, Nawab, et al., 2019). Inflammatory or homeostatic cytokines include low molecular weight (LMW) proteins (8–14 kDa) with 47 human receptors. These are further subdivided into four family members, based on the location of the two cysteine residues located at the amino terminus (CXC, CC, XC, CX3C) (Gao & Fish, 2018).

The 19 unique GPCRs interact with 47 distinct chemokines. The cognate GPCR undergoes a conformational change when the ligand binds, dissociating the Gαi and Gβγ subunits and thereby inducing signaling. Specifically, Gβγ subunits activate phospholipase Cγ, PIP2, and IP3, leading to a rapid increase in cytosolic Ca^{+2}. Previous literature has shown that CCR5 activation of Ca^{+2} signaling and cellular migration is preserved in immunologic (Olson et al., 1999) and malignant cells (Jiao et al., 2018; Velasco-Velazquez et al., 2012). The induction of additional pathways by CCR5, include the PI-3′K pathway and thereby PDK1 and serine/threonine kinase protein kinase B (AKT), which in turn induces cell survival, glycolysis, cell proliferation, growth and proliferation of progenitor and stem cells, immune cell differentiation and the release of eIF4E to promote cap-dependent translation (Fig. 1A).

CCR5 serves as a lynchpin in the function of T-helper cells, macrophages, eosinophils, myeloid derived suppressor cells (MDSC), microglia and dendritic cells (Fig. 1B). In addition, CCR5 can be pathologically expressed during cellular transformation. The induction of CCR5 expression leads cells to changes in genomic characteristics as well as its metastatic behavior.

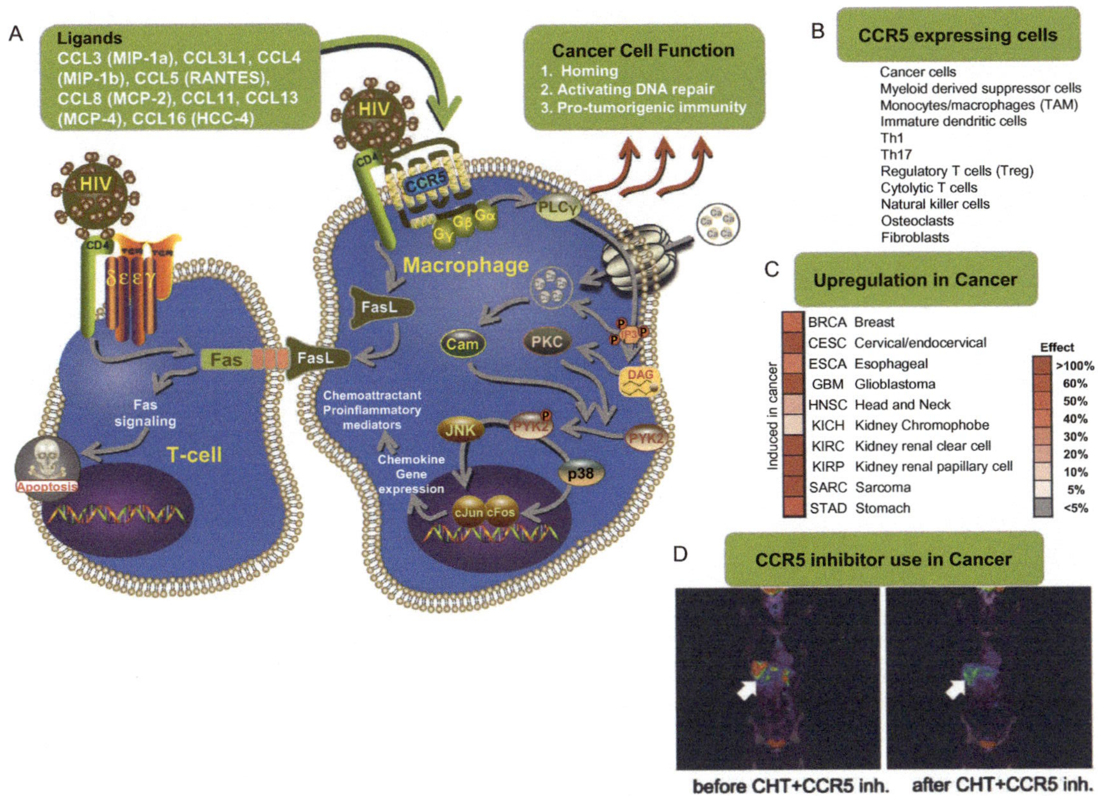

Fig. 1 See legend on opposite page.

3. CCR5 antagonists retasked in cancer (Jiao et al., 2018)

Based on the importance of CCR5 activation in the onset and progression of cancer, several CCR5 antagonists that were developed for HIV therapy, are now being retested for cancer and cancer-related diseases. The two primary CCR5 inhibitors, maraviroc and vicriviroc were used in clinical trials to treat patients with HIV. Leronlimab, a humanized monoclonal antibody inhibitor, has also demonstrated therapeutic benefit in HIV patients and has been used in >760 HIV^+ patients without causing any severe therapy-related adverse events. Leronlimab and Maraviroc reached their primary efficacy endpoints in Phase 3 HIV clinical trials (Dhody et al., 2018; Fatkenheuer et al., 2008; Kaplon & Reichert, 2019, 2018).

In a mouse model of pancreatic adenocarcinoma, TAK-779, a quaternary ammonium derivative, reduced T regulatory cell infiltration and tumor growth (Tan et al., 2009). Anibamine, a natural product that antagonizes CCR5, decreased the growth, adhesion, and invasion of prostate cancer. In other murine cancer models, Met-CCL5, a competitive chemokine receptor antagonist, decreased not only the tumor growth of malignant breast cancer cells, but also the infiltration of macrophages. Aplaviroc, a

Fig. 1 CCR5 signaling in immune and cancer cells. (A) Schematic representation of a T cell expressing CCR5 with the intracellular signaling cascade activated by ligand. The diverse ligands for CCR5 are shown in green. The pathological response induced by CCR5 on cancer cells is shown in the yellow box. (B) The diverse type of cells expressing CCR5 are shown. (C) CCR5 expression derived from TCGA with square indicating effect of upregulation in cancer vs normal tissue shown as a calorimetric display of fold increase in expression as a hazards ratio. (RNA-seq TCGA data (v2 RSEM values) was downloaded using Fire-browse and FPKM values were log2-scaled and quantile normalized (Mean differences between cancer and normal tissues were calculated). (D) Representative PET-MRI images from a patient receiving chemotherapy (CHT) after participation in the phase 1 pilot MARACON study, in which patients with advanced-stage metastatic colorectal cancer who were refractory to standard chemotherapy (Halama et al., 2016), were treated with maraviroc. White arrow indicates liver with metastatic lesions. Red spots indicate high glucose uptake typical for metastases, and green indicates low background glucose uptake. *Adopted from Jiao, X., Nawab, O., Patel, T., Kossenkov, A. V., Halama, N., Jaeger, D., et al. (2019). Recent advances targeting CCR5 for cancer and its role in immuno-oncology.* Cancer Research, 79, *4801–4807. https://doi.org/10.1158/0008-5472.CAN-19-1167.*

2,5-diketopiperazine CCR5 entry inhibitor, has been discontinued due to its hepatotoxic side effect profile (Nichols et al., 2008). A reduction in CCR5 expression and cancer cell migration was achieved via a saponin named DT-13. The utilization of zinc finger nuclease and siRNA have also demonstrated efficacy in reducing CCR5 expression. GAG mimetics that inhibit CCL5 binding to GAG, OTR4120 and OTR4131, minimize CCL5-induced invasion and migration of hepatocellular carcinoma (HCC). Because of its failed efficacy and side effect profile, INCB9471 (Incyte Corporation) was discontinued after a phase 2 trial for HIV positive patients (Shin et al., 2011). Cinicriviroc (TBR-652) (Takeda), which is a dual CCR2-CCR5 inhibitor has completed a Phase IIb clinical trial for HIV and is now being researched for its utility in treating Nonalcoholic Steatohepatitis (NASH). Also, a soluble receptor-based fusion protein named mCCR5-Ig was found to inhibit CCR5 and is being further investigated for its therapeutic potential (Ryu et al., 2018; Sugasawa, Ichikura, Tsujimoto, et al., 2008).

CCR5 overexpression has been demonstrated in breast (Jiao et al., 2018; Velasco-Velazquez et al., 2012), prostate (Sicoli et al., 2014), and colorectal cancer (Halama et al., 2016; Pervaiz et al., 2015), as well as melanoma (Liu et al., 2018), lymphoma (Casagrande et al., 2018), head and neck cancer (Gonzalez-Arriagada et al., 2018), gastric cancer (Aldinucci & Casagrande, 2018), esophageal cancer (Wu et al., 2018), and pancreatic carcinoma (Fig. 1B). In the analysis of >2200 breast cancer patients, >50% of patients' tumors were $CCR5^+$ and >95% of triple negative breast cancer (TNBC) were $CCR5^+$ (Velasco-Velazquez et al., 2012, 2014). It was found that the higher the cytoplasmic CCR5 staining, the poorer was the prognosis (Jiao, Nawab, et al., 2019; Jiao, Wang, & Pestell, 2019). CCR5 is induced by oncogenic transformation (Ha-Ras, c-Myc, ErbB2, c-Src) (Velasco-Velazquez et al., 2012), DNA damage (Jiao et al., 2018; Jiao, Wang, & Pestell, 2019) and CCL5. CCR5 receptor levels correspond to poor prognosis in breast cancer and gastric adenocarcinoma (Jiao et al., 2018; Ryu et al., 2018; Sugasawa, Ichikura, Tsujimoto, et al., 2008).

Although CCR5 binds to many ligands in cellular proliferation, elevated levels of the ligand indicates a poor prognosis in breast (Niwa et al., 2001; Yaal-Hahoshen et al., 2006), cervical (Niwa et al., 2001), prostatic (Tsukishiro et al., 2006), gastric (Sima et al., 2014), (Sugasawa, Ichikura, Tsujimoto, et al., 2008), colorectal (Suenaga et al., 2016) and pancreatic cancers. (Cambien et al., 2011; Sugasawa, Ichikura, Kinoshita, et al., 2008; Vangelista & Vento, 2017).

4. CCR5 induces the hallmarks of cancer

CCR5 expression induces cancer cell homing behavior to metastatic sites (Velasco-Velazquez et al., 2012, 2014; Zi et al., 2017). CCR5 alters the pro-inflammatory metastatic immune phenotype (Halama et al., 2016) and enhances DNA repair, rendering cells resistant to DNA damaging agents (Jiao et al., 2018; Jiao, Nawab, et al., 2019; Jiao, Wang, & Pestell, 2019).

4.1 Activating invasion and metastasis (Jiao et al., 2018; Velasco-Velazquez et al., 2012)

Unique, distinct mechanisms govern the pathophysiology of tumor invasion vs. metastasis (Liu et al., 2014; Massague & Obenauf, 2016). Ectopic CCR5 expression within apical epithelial cells is sufficient to drive cancer cell metastasis (Velasco-Velazquez et al., 2012). Small molecular inhibitors specific to CCR5 also block the metastasis of isogenic oncogene-transformed breast cancer cells in NOD/SCID mice (Velasco-Velazquez et al., 2012) and prostate cancer metastasis in immunocompetent mice (Sicoli et al., 2014). Maraviroc inhibits SUP-B15 cells migration to CXCL12 and CXCL13, in vitro (Zi et al., 2017). CCR5 induces metastasis in $p53^{-}$ breast cancer cells in vivo (Jiao et al., 2018; Zhang et al., 2018).

4.2 Avoiding immune destruction-the anti-tumor immune response (Blattner et al., 2018; Hawila et al., 2017)

Tumorigenesis induces ligands for CCR5 promoting pro-metastatic and inflammatory pathways via distinct mechanisms. Recent evidence supports a hypothesis that CCR5 inhibitors may have a potential synergy with the canonical immune checkpoint inhibitors. In current clinical trials by Pfizer and Merck, CCR5 inhibitors (Maraviroc, Vicriviroc) are being used in combination with a checkpoint inhibitor (Pembrolizumab). The recruitment and migration of innate lymphoid cells (ILCs) tumor-infiltrating lymphocytes (TILs), Tregs (Schlecker et al., 2012) mesenchymal stem cells (MSCs) and immature dendritic cells (DCs), participate in tumor-induction and anti-tumor immunosuppression (Sleeman, 2015). The mechanism by which tumors evade destruction involves an active process of inducing immune tolerance via recruitment of $CD4^{+}CD25^{+}Foxp3^{+}$ regulatory T cells (Treg).

Cellular expression of CCR5 occurs in many other cell types including stem cells (Fig. 1). For instance, MSC produce CCL3, CCCL4 and CCL5.

Maraviroc reduces the MDSC-induction of colon cancer metastasis (Nishikawa et al., 2019). Furthermore, $CD4^+$ $Foxp3^+$ Treg, preferentially express CCR5 compared with $CD4^+$ $Foxp3^-$ effector T cells; which participate in the migration and growth of pancreatic adenocarcinoma (Tan et al., 2009).

Conversely, an absence of CCR5 ligand expression has a strong correlation to a diminished infiltration of antigen specific T cells and metastasis (Harlin et al., 2009). It is also been hypothesized that through TGFβ (Chang, Lin, Mahalingam, et al., 2012), tumor-derived CCL5 impedes the anti-tumor T cell response to thereby enhance proliferation of murine mammary carcinoma (Adler et al., 2003). CCR5 is part of a CCL3-CCR5/CCR1-mediated DC cell migration to lymph nodes and the tumor microenvironment (TME). $CD4^+$ T cells interact with DC, CCL3, and CCL4 which are released into tissues for the activation of $CCR5^+$ naïve $CD8^+$Tcells (Castellino et al., 2006). CCR5 and its ligands promote the proliferation of $CCR5^+$ PMN-MDSCs into the bone marrow and, later, trigger immunosuppressive activities at the site of tumor burden via arginase-1.

CCR5 guides $CD11b^+Gr1^+Ly6C^{low}$ polymorphonuclear myeloid cells from the bone marrow to promote tumor development (Hawila et al., 2017). The subtypes $CD11b^+Ly6G^-Ly6C^{hi}$ monocytic MDSCs and $CD11b^+Ly6G^+Ly6C^{low}$ polymorphonuclear (PMN) both contribute to tumor growth and immunosuppression (Hawila et al., 2017). In mice, CCR5 blockade with anti-CCR5 antibody decreases growth of B16 melanoma and reduces MDSC accumulation. Also, the endogenous ligand CCL8, is produced by $F4/80^+$ macrophages in the lungs of mice with metastatic primary tumors (Halvorsen et al., 2016). Reduction in the migration of Tregs towards CCL8 ex vivo in the presence of the CCR5 inhibitor Maraviroc has been demonstrated. Finally, mice that were treated with Maraviroc showed reduction in the level of $CCR5^+$ Tregs and metastatic tumor burden in the lungs (Halvorsen et al., 2016).

TAMs express CCR5; they are comprised of macrophages (M1–M2 spectrum) and express variable levels of arginase and interleukins (IL4, IL10, and IL13). $F4/80^+$ macrophages are strongly implicated in human cancer progression and are well known participants in the onset and progression of mammary tumors in murine models (Qian & Pollard, 2010). Recruitment of TAMs toward the tumor burden are facilitated by ligands including RANTES (Azenshtein et al., 2002). Through binding to CCR5 and CCR1, CCL3 promotes tumorigenesis through recruitment of pro-tumorigenic macrophages into the TME (Kitamura et al., 2015). Interestingly, gene deletion of *CCL3* in murine macrophages reduced the number of lung metastasis and an adoptive transfer of wild type inflammatory monocytes increased the number of lung metastasis in *Ccl3* deficient mice (Kitamura et al., 2015).

Other ligands including EGF, CSF1, HGF, CCL2, CXCR4/CXCl12 and Tie2 collectively contribute to the diversity of inflammatory subtypes seen within the tumor microenvironment necessary for tumor progression (Arwert et al., 2018). Moreover, single cell sequencing to assess the expansion of immune cell phenotypes and diversity of cell states within the tumor microenvironment demonstrated that there exists evidence for *CCR5* as a gene that is highly correlated with the activation of tumor heterogeneity found in the breast cancer tumor microenvironment (Azizi et al., 2018).

Moreover, several lines of evidence suggest that CCR5 and its ligands appear to participate in the canonical immune check point response. The Programmed Cell Death Protein 1 (CD279 and PD-1) and its ligand PD-1 Ligand (PD-L1) signaling pathway are critical players in immune checkpoints. PD-1 signaling plays an important role in tumor evasion from the innate immune system. In addition, tumor-infiltrating lymphocytes (TILs) serve as a biomarker for predicting responses to PD-L1 blockade therapy. Recent evidence supports that tumors responsive to immunotherapy tend to infiltrate with T cells in response to CTLA-4 and PD-1 antagonism; this process is referred to as a "T cell-inflamed" tumor microenvironment (Gajewski, 2015; Ji et al., 2012; Tumeh et al., 2014). CCL5 upregulation was demonstrated in PD-L1-positive melanoma tumors along with IFNγ and several IFNγ-related genes (Ayers et al., 2017; Taube et al., 2015). Tumor burden and a T cell-inflammatory gene expression were independently predictive of the response to the PD-1 antibody pembrolizumab (Cristescu et al., 2018).

It is also important to analyze the role of MDSC. $CCR5^{high}$ MDSC have higher immunosuppressive activity than $CCR5^{low}$ MDSC (Chang, Lin, Kang, et al., 2012), and when disrupted, MDSC trafficking enhances anti-PD1 therapy (Highfill et al., 2014). As described above, CCL5 promotes not only the influx of $CD8^+$ T-cells (Harlin et al., 2009), but also the expression of PD-L1 when associated with increased TILs. Relatedly, the Keynote-028 study demonstrated that the tumors with high PD-L1 and T cell inflammatory gene expression, were associated with a positive outcome from monoclonal antibody therapy (pembrolizumab) (Seto et al., 2019). Collectively, these studies suggest that overlap exists between the CCR5 non-canonical and canonical immune checkpoint pathways.

4.3 Induction of proliferative signaling, angiogenesis (Ben-Baruch, 2012; Soria & Ben-Baruch, 2008) and resistance to cell death (Jiao, Nawab, et al., 2019; Jiao, Wang, & Pestell, 2019)

Support for the requirement of CCR5 in oncogene-induced cellular proliferation was provided by transgenic studies in which MMTV-PyMT-induced mammary tumors were reduced in $CCR5^{-/-}$ mice (Gao et al., 2017). CCL5 promotes cellular spreading and tumor migration. Malignant cells produce VEGF upon CCL5 stimulation and by secreting CCL5 recruit CCR5-expressing TAMs (Frankenberger et al., 2015; Robinson et al., 2003). There is also evidence that CCR5 inhibitors can reduce angiogenesis in triple negative breast cancer (TNBC) cell line xenografts (Kang et al., 2009; Wang et al., 2016).

4.4 Deregulated cellular energetics and cancer stem cells

For cells to undergo uncontrolled division, they require an environment that provides them increased amounts of glucose uptake (Gao & Fish, 2018; Gao, Rahbar, & Fish, 2016; Martinez-Outschoorn, Peiris-Pages, Pestell, Sotgia, & Lisanti, 2016). Analysis of MDA-MB-231 breast cancer cell line demonstrated that CCR5 governs a dramatic (>1000-fold) increase in activation of RNA governing ribosomal biogenesis and cell survival signaling pathways (Jiao et al., 2018). Furthermore, $CCR5^+$ breast cancer epithelial cells revealed features of oncogenic stem cells; forming mammospheres and initiating tumorigenesis with a >60-fold efficiency in mice (Jiao et al., 2018; Jiao, Wang, & Pestell, 2019).

5. Preclinical analysis of CCR5 inhibition in metastatic cancer

CCR5 antagonists (maraviroc and vicriviroc) blocked metastasis of human breast cancer xenografts (MDA-MB-231 cells) in immuno-deficient mice. This process was mediated by the inhibition of the tumor cell homing as well as the enhanced killing of these cells via DNA damaging agents (Jiao et al., 2018; Jiao, Wang, & Pestell, 2019; Velasco-Velazquez et al., 2012). The combination of maraviroc along with nanotechnology targeting CCL5 in the bone marrow augmented anti-tumor immunity (Ban et al., 2017). Both maraviroc and vicriviroc reduced prostate cancer cell metastasis to bone and brain in immune-competent mice (Sicoli et al., 2014). Partly by limiting fibroblast accumulation, maraviroc reduced the growth and spread of colon cancer in orthotopically injected mice (Tanabe et al., 2016). In addition, anti-CCR5 antibody inhibited melanoma (B16) cellular proliferation and MDSC accumulation in tumor tissues (Tang et al., 2015). TAK-779 reduced pancreatic cancer cell growth and metastasis with a noticeable reduction in the migration of Tregs into the tumors (Tan et al., 2009).

Evidence has also suggested that chemokines participate in the development of nonalcoholic steatohepatitis (NASH) and thereby progression to hepatocellular cancer (Kutlu et al., 2018). In this context, maraviroc reduced lipogenesis and β-oxidation in NASH, thereby decreasing insulin resistance and steatosis. Maraviroc therapy decreased the development of NASH and hepatocellular carcinoma in a murine model of NASH (Ochoa-Callejero et al., 2013; Perez-Martinez et al., 2018).

6. Clinical studies

Patients with advanced-stage metastatic colorectal cancer (CRC), who were refractory to standard chemotherapy (Halama et al., 2016), were treated with maraviroc (The objective response rates in metastatic CRC patients on, or after, the third line of chemotherapy, is typically around 5%–10% to other agents). In this phase 1 pilot (MARACON) study, a reduced proliferation rate was seen in all tumor samples examined. An inverse correlation was also found between "immune CCR5" levels and the maturation status of tumor-infiltrating neutrophils as well as 5-year-survival rates (Ban et al., 2017). From the 11 patients of the core cohort, an additional five patients were given further chemotherapy. Three of these five patients had objective partial responses. These findings compared favorably with historical response rates in this patient population.

Three additional studies which combine a drug and a biologic for $CCR5^{+}$ metastatic cancer have been approved by the FDA. The phase 1 study of pembrolizumab with maraviroc in patients with refractory microsatellite stable (MSS) CRC is supported by Pfizer (Table 1). The second, a phase 2 study is assessing safety and efficacy of vicriviroc in combination with pembrolizumab (MK-3475) in patients with advanced metastatic MSS-CRC. The third is a phase 1b/2 study using carboplatin and leronlimab for $CCR5^{+}$ metastatic triple negative breast cancer (Table 1). The impact on progression-free survival (PFS) will be assessed. Interestingly, the secondary objectives will assess both the overall response rate (ORR), the number of circulating tumor cells (CTC), and will measure

Table 1 Active clinical trials using CCR5 inhibitors.

CCR5 Antagonist	Mechanism	Administration	Reference
Maraviroc UK-427857 (Pfizer) approved for HIV 2007		Maraviroc (days 1–21 of each cycle) + Pembrolizumab (day 1, day 22)	1. Phase 1 study of Pembrolizumab with Maraviroc in patients with refractory microsatellite stable colorectal cancer. NCT03274804. 2. GVHD, Phase 1 NCT00948753. Phase 2 NCT02167451.
Vicriviroc CH 417690 (Merck)	Pyrimidine CCR5 inhibitor	Vicriviroc + Pembrolizumab	1. Phase 2, Vicriviroc in combination with Pembrolizumab (MK-3475) in patients with advanced metastatic microsatellite stable colorectal cancer. NCT03631407.
Leronlimab (Pro-140) (CytoDyn)	Humanized monoclonal antibody	Weekly self-injection	1. Phase 2 study for $CCR5^+$ triple negative breast cancer using Carboplatin and Leronlimab NCT03838367. 2. Phase 2 GVHD NCT02737306.

time to new metastasis. As CTC express CCR5 and the number of CTC reflect metastatic tumor burden, the outcome of CCR5 inhibitors on $CCR5^+$ CTC my provide a useful surrogate measurement in the future for metastatic cancer studies.

In studies of hepatocellular cancer prevention targeting NASH, liver fibrosis improved after 1 year of therapy with cenicriviroc, leading to the implementation of a phase 3 trial (AURORA) (Tacke, 2018). Tropifexor (LJN452) and cenicriviroc are being assessed for safety, tolerability, and efficacy in patients with NASH and liver fibrosis (TANDEM) (NCT03517540).

Relevant progress has been made in the treatment of bone marrow transplant-related graft vs. host disease (GVHD) with CCR5 inhibitors. A trial has shown maraviroc decreased the incidence of GVHD without incurring a likelihood of disease relapse (Moy et al., 2017). Maraviroc is therefore being assessed for GVHD prophylaxis in pediatric and adult patients who require stem cell transplants (NCT02167451). Based on the reduction of GVHD in a murine model (Burger et al., 2018) an open-label, single-arm, Phase II study of the safety and efficacy of leronlimab (Pro-140) is being conducted for prophylaxis of GVHD in patients undergoing reduced intensity conditioning (RIC) allogeneic stem-cell transplantation (NCT02737306).

The GPCR CCR5 has become an important new target for cancer therapies because of the clear role in metastasis and the anti-tumor immune response of several cancers. The current clinical trials have been open to accrual for less than a year at the time of writing this review. Chemotherapy and radiation induced side effects remain substantial. Given the increased selective cancer cell killing that occurs when CCR5 inhibitors are added to either chemotherapy or radiation (Jiao et al., 2018) the potential for reducing the dose and thereby side effects of current cancer therapies through deploying CCR5 inhibitors represents a substantial opportunity to improve patients' quality of life. Given the importance of CCR5 in HIV (Dean et al., 1996; Gupta et al., 2019; Samson et al., 1996), progression of stroke (Joy et al., 2019), and osteoporosis (Xie, Zhan, Ge, & Tang, 2019), the clinical knowledge learned from cancer therapies may have important cross disciplinary value for other human diseases.

Acknowledgments

This work was supported in part by Breakthrough Breast Cancer Research Program grant award # W81XWH1810605 and by R01CA132115 (R.G.P).

Conflict of interest

R.G.P. holds ownership interests in the biopharmaceutical company CytoDyn and holds ownership interests in LightSeed, Inc. R.G.P. additionally holds ownership interests (value unknown) for several patents and submitted patent applications. The other authors declare that they have no conflict of interest.

References

Adler, E. P., Lemken, C. A., Katchen, N. S., & Kurt, R. A. (2003). A dual role for tumor-derived chemokine RANTES (CCL5). *Immunology Letters*, *90*(2–3), 187–194. Retrieved from https://www.ncbi.nlm.nih.gov/pubmed/14687724.

Aldinucci, D., & Casagrande, N. (2018). Inhibition of the CCL5/CCR5 axis against the progression of gastric cancer. *International Journal of Molecular Sciences*, *19*(5), 1477. https://doi.org/10.3390/ijms19051477.

Alonzo, F., 3rd, Kozhaya, L., Rawlings, S. A., Reyes-Robles, T., DuMont, A. L., Myszka, D. G., et al. (2013). CCR5 is a receptor for *Staphylococcus aureus* leukotoxin ED. *Nature, 493*(7430), 51–55. https://doi.org/10.1038/nature11724.

Arwert, E. N., Harney, A. S., Entenberg, D., Wang, Y., Sahai, E., Pollard, J. W., et al. (2018). A unidirectional transition from migratory to perivascular macrophage is required for tumor cell intravasation. *Cell Reports, 23*(5), 1239–1248. https://doi.org/10.1016/j.celrep.2018.04.007.

Ayers, M., Lunceford, J., Nebozhyn, M., Murphy, E., Loboda, A., Kaufman, D. R., et al. (2017). IFN-gamma-related mRNA profile predicts clinical response to PD-1 blockade. *The Journal of Clinical Investigation, 127*(8), 2930–2940. https://doi.org/10.1172/JCI91190.

Azenshtein, E., Luboshits, G., Shina, S., Neumark, E., Shahbazian, D., Weil, M., et al. (2002). The CC chemokine RANTES in breast carcinoma progression: Regulation of expression and potential mechanisms of promalignant activity. *Cancer Research, 62*(4), 1093–1102. Retrieved from https://www.ncbi.nlm.nih.gov/pubmed/11861388.

Azizi, E., Carr, A. J., Plitas, G., Cornish, A. E., Konopacki, C., Prabhakaran, S., et al. (2018). Single-cell map of diverse immune phenotypes in the breast tumor microenvironment. *Cell, 174*(5), 1293–1308.e1236. https://doi.org/10.1016/j.cell.2018.05.060.

Ban, Y., Mai, J., Li, X., Mitchell-Flack, M., Zhang, T., Zhang, L., et al. (2017). Targeting autocrine CCL5-CCR5 axis reprograms immunosuppressive myeloid cells and reinvigorates antitumor immunity. *Cancer Research,* 77(11), 2857–2868. https://doi.org/10.1158/0008-5472.CAN-16-2913.

Ben-Baruch, A. (2012). The tumor-promoting flow of cells into, within and out of the tumor site: Regulation by the inflammatory axis of TNFalpha and chemokines. *Cancer Microenvironment, 5*(2), 151–164. https://doi.org/10.1007/s12307-011-0094-3.

Blanpain, C., Lee, B., Tackoen, M., Puffer, B., Boom, A., Libert, F., et al. (2000). Multiple nonfunctional alleles of CCR5 are frequent in various human populations. *Blood, 96*(5), 1638–1645. Retrieved from https://www.ncbi.nlm.nih.gov/pubmed/10961858.

Blattner, C., Fleming, V., Weber, R., Himmelhan, B., Altevogt, P., Gebhardt, C., et al. (2018). CCR5(+) myeloid-derived suppressor cells are enriched and activated in melanoma lesions. *Cancer Research, 78*(1), 157–167. https://doi.org/10.1158/0008-5472.CAN-17-0348.

Burger, D. R., Parker, Y., Guinta, K., & Lindner, D. (2018). PRO 140 monoclonal antibody to CCR5 prevents acute xenogeneic graft-versus-host disease in NOD-scid IL-2Ry(null) mice. *Biology of Blood and Marrow Transplantation, 24*(2), 260–266. https://doi.org/10.1016/j.bbmt.2017.10.041.

Cambien, B., Richard-Fiardo, P., Karimdjee, B. F., Martini, V., Ferrua, B., Pitard, B., et al. (2011). CCL5 neutralization restricts cancer growth and potentiates the targeting of PDGFRbeta in colorectal carcinoma. *PLoS One, 6*(12), e28842. https://doi.org/10.1371/journal.pone.0028842.

Casagrande, N., Borghese, C., Visser, L., Mongiat, M., Colombatti, A., & Aldinucci, D. (2018). CCR5 antagonism by maraviroc inhibits Hodgkin lymphoma microenvironment interactions and xenograft growth. *Haematologica, 104,* 564–575. https://doi.org/10.3324/haematol.2018.196725.

Castellino, F., Huang, A. Y., Altan-Bonnet, G., Stoll, S., Scheinecker, C., & Germain, R. N. (2006). Chemokines enhance immunity by guiding naive CD8+ T cells to sites of CD4+ T cell-dendritic cell interaction. *Nature, 440*(7086), 890–895. https://doi.org/10.1038/nature04651.

Chang, L. Y., Lin, Y. C., Kang, C. W., Hsu, C. Y., Chu, Y. Y., Huang, C. T., et al. (2012). The indispensable role of CCR5 for in vivo suppressor function of tumor-derived CD103+ effector/memory regulatory T cells. *Journal of Immunology, 189*(2), 567–574. https://doi.org/10.4049/jimmunol.1200266.

Chang, L. Y., Lin, Y. C., Mahalingam, J., Huang, C. T., Chen, T. W., Kang, C. W., et al. (2012). Tumor-derived chemokine CCL5 enhances TGF-beta-mediated killing of CD8(+) T cells in colon cancer by T-regulatory cells. *Cancer Research*, *72*(5), 1092–1102. https://doi.org/10.1158/0008-5472.CAN-11-2493.

Cristescu, R., Mogg, R., Ayers, M., Albright, A., Murphy, E., Yearley, J., et al. (2018). Pan-tumor genomic biomarkers for PD-1 checkpoint blockade-based immunotherapy. *Science*, *362*(6411), eaar3593. https://doi.org/10.1126/science.aar3593.

Dean, M., Carrington, M., Winkler, C., Huttley, G. A., Smith, M. W., Allikmets, R., et al. (1996). Genetic restriction of HIV-1 infection and progression to AIDS by a deletion allele of the CKR5 structural gene. Hemophilia Growth and Development Study, Multicenter AIDS Cohort Study, Multicenter Hemophilia Cohort Study, San Francisco City Cohort, ALIVE Study. *Science*, *273*(5283), 1856–1862. https://doi.org/10.1126/science.273.5283.1856.

Dhody, K., Kazempour, K., Pourhassan, N., & Maddon, P. J. (2018). Primary efficacy results of PRO 140 SC in a pivotal phase 2b/3 study in heavily treatment-experienced HIV-1 patients. In *Paper presented at the ASM/ICAAC.* Atlanta, Georgia: ASM/ICAAC. June 7–11.

Fatkenheuer, G., Nelson, M., Lazzarin, A., Konourina, I., Hoepelman, A. I., Lampiris, H., et al. (2008). Subgroup analyses of maraviroc in previously treated R5 HIV-1 infection. *The New England Journal of Medicine*, *359*(14), 1442–1455. https://doi.org/10.1056/NEJMoa0803154.

Frankenberger, C., Rabe, D., Bainer, R., Sankarasharma, D., Chada, K., Krausz, T., et al. (2015). Metastasis suppressors regulate the tumor microenvironment by blocking recruitment of prometastatic tumor-associated macrophages. *Cancer Research*, *75*(19), 4063–4073. https://doi.org/10.1158/0008-5472.CAN-14-3394.

Gajewski, T. F. (2015). The next hurdle in cancer immunotherapy: Overcoming the non-T-cell-inflamed tumor microenvironment. *Seminars in Oncology*, *42*(4), 663–671. https://doi.org/10.1053/j.seminoncol.2015.05.011.

Gao, D., Cazares, L. H., & Fish, E. N. (2017). CCL5-CCR5 interactions modulate metabolic events during tumor onset to promote tumorigenesis. *BMC Cancer*, *17*(1), 834. https://doi.org/10.1186/s12885-017-3817-0.

Gao, D., & Fish, E. N. (2018). Chemokines in breast cancer: Regulating metabolism. *Cytokine*, *109*, 57–64. https://doi.org/10.1016/j.cyto.2018.02.010.

Gao, D., Rahbar, R., & Fish, E. N. (2016). CCL5 activation of CCR5 regulates cell metabolism to enhance proliferation of breast cancer cells. *Open Biology*, *6*(6), 160122. https://doi.org/10.1098/rsob.160122.

Gonzalez-Arriagada, W. A., Lozano-Burgos, C., Zuniga-Moreta, R., Gonzalez-Diaz, P., & Coletta, R. D. (2018). Clinicopathological significance of chemokine receptor (CCR1, CCR3, CCR4, CCR5, CCR7 and CXCR4) expression in head and neck squamous cell carcinomas. *Journal of Oral Pathology & Medicine*, *47*(8), 755–763. https://doi.org/10.1111/jop.12736.

Gupta, R. K., Abdul-Jawad, S., McCoy, L. E., Mok, H. P., Peppa, D., Salgado, M., et al. (2019). HIV-1 remission following CCR5Delta32/Delta32 haematopoietic stem-cell transplantation. *Nature*, *568*(7751), 244–248. https://doi.org/10.1038/s41586-019-1027-4.

Halama, N., Zoernig, I., Berthel, A., Kahlert, C., Klupp, F., Suarez-Carmona, M., et al. (2016). Tumoral immune cell exploitation in colorectal cancer metastases can be targeted effectively by Anti-CCR5 therapy in cancer patients. *Cancer Cell*, *29*(4), 587–601. https://doi.org/10.1016/j.ccell.2016.03.005.

Halvorsen, E. C., Hamilton, M. J., Young, A., Wadsworth, B. J., LePard, N. E., Lee, H. N., et al. (2016). Maraviroc decreases CCL8-mediated migration of CCR5(+) regulatory T cells and reduces metastatic tumor growth in the lungs. *Oncoimmunology*, *5*(6), e1150398. https://doi.org/10.1080/2162402X.2016.1150398.

Harlin, H., Meng, Y., Peterson, A. C., Zha, Y., Tretiakova, M., Slingluff, C., et al. (2009). Chemokine expression in melanoma metastases associated with CD8+ T-cell recruitment. *Cancer Research, 69*(7), 3077–3085. https://doi.org/10.1158/0008-5472.CAN-08-2281.

Hawila, E., Razon, H., Wildbaum, G., Blattner, C., Sapir, Y., Shaked, Y., et al. (2017). CCR5 directs the mobilization of CD11b(+)Gr1(+)Ly6C(low) polymorphonuclear myeloid cells from the bone marrow to the blood to support tumor development. *Cell Reports, 21*(8), 2212–2222. https://doi.org/10.1016/j.celrep.2017.10.104.

Highfill, S. L., Cui, Y., Giles, A. J., Smith, J. P., Zhang, H., Morse, E., et al. (2014). Disruption of CXCR2-mediated MDSC tumor trafficking enhances anti-PD1 efficacy. *Science Translational Medicine, 6*(237), 237ra267. https://doi.org/10.1126/scitranslmed.3007974.

Ji, R. R., Chasalow, S. D., Wang, L., Hamid, O., Schmidt, H., Cogswell, J., et al. (2012). An immune-active tumor microenvironment favors clinical response to ipilimumab. *Cancer Immunology, Immunotherapy, 61*(7), 1019–1031. https://doi.org/10.1007/s00262-011-1172-6.

Jiao, X., Nawab, O., Patel, T., Kossenkov, A. V., Halama, N., Jaeger, D., et al. (2019). Recent advances targeting CCR5 for cancer and its role in immuno-oncology. *Cancer Research, 79*, 4801–4807. https://doi.org/10.1158/0008-5472.CAN-19-1167.

Jiao, X., Velasco-Velazquez, M. A., Wang, M., Li, Z., Rui, H., Peck, A. R., et al. (2018). CCR5 governs DNA damage repair and breast cancer stem cell expansion. *Cancer Research, 78*(7), 1657–1671. https://doi.org/10.1158/0008-5472.CAN-17-0915.

Jiao, X., Wang, M., & Pestell, R. G. (2009). Leronlimab, a humanized monoclonal antibody to CCR5, blocks breast cancer cellular invasion and enhances cell death induced by DNA damaging chemotherapies. In *AACR Annual Meeting 2019; March 29–April 3, 2019; Atlanta, GA*, Abstract. https://doi.org/10.1158/1538-7445.AM2019-2009.

Joy, M. T., Ben Assayag, E., Shabashov-Stone, D., Liraz-Zaltsman, S., Mazzitelli, J., Arenas, M., et al. (2019). CCR5 is a therapeutic target for recovery after stroke and traumatic brain injury. *Cell, 176*(5), 1143–1157.e1113. https://doi.org/10.1016/j.cell.2019.01.044.

Kang, S., Lee, S. P., Kim, K. E., Kim, H. Z., Memet, S., & Koh, G. Y. (2009). Toll-like receptor 4 in lymphatic endothelial cells contributes to LPS-induced lymphangiogenesis by chemotactic recruitment of macrophages. *Blood, 113*(11), 2605–2613. https://doi.org/10.1182/blood-2008-07-166934.

Kaplon, H., & Reichert, J. M. (2018). Antibodies to watch in 2018. *MAbs, 10*(2), 183–203. https://doi.org/10.1080/19420862.2018.1415671.

Kaplon, H., & Reichert, J. M. (2019). Antibodies to watch in 2019. *MAbs, 11*(2), 219–238. https://doi.org/10.1080/19420862.2018.1556465.

Kindberg, E., Mickiene, A., Ax, C., Akerlind, B., Vene, S., Lindquist, L., et al. (2008). A deletion in the chemokine receptor 5 (CCR5) gene is associated with tickborne encephalitis. *The Journal of Infectious Diseases, 197*(2), 266–269. https://doi.org/10.1086/524709.

Kitamura, T., Qian, B. Z., Soong, D., Cassetta, L., Noy, R., Sugano, G., et al. (2015). CCL2-induced chemokine cascade promotes breast cancer metastasis by enhancing retention of metastasis-associated macrophages. *The Journal of Experimental Medicine, 212*(7), 1043–1059. https://doi.org/10.1084/jem.20141836.

Kutlu, O., Kaleli, H. N., & Ozer, E. (2018). Molecular pathogenesis of nonalcoholic steatohepatitis– (NASH–) related hepatocellular carcinoma. *Canadian Journal of Gastroenterology & Hepatology, 2018*, 8543763. https://doi.org/10.1155/2018/8543763.

Lalani, A. S., Masters, J., Zeng, W., Barrett, J., Pannu, R., Everett, H., et al. (1999). Use of chemokine receptors by poxviruses. *Science, 286*(5446), 1968–1971. https://doi.org/10.1126/science.286.5446.1968.

Lim, J. K., Louie, C. Y., Glaser, C., Jean, C., Johnson, B., Johnson, H., et al. (2008). Genetic deficiency of chemokine receptor CCR5 is a strong risk factor for symptomatic West Nile virus infection: A meta-analysis of 4 cohorts in the US epidemic. *The Journal of Infectious Diseases*, *197*(2), 262–265. https://doi.org/10.1086/524691.
Liu, W., Vivian, C. J., Brinker, A. E., Hampton, K. R., Lianidou, E., & Welch, D. R. (2014). Microenvironmental influences on metastasis suppressor expression and function during a metastatic cell's journey. *Cancer Microenvironment*, 7(3), 117–131. https://doi.org/10.1007/s12307-014-0148-4.
Liu, J., Wang, C., Ma, X., Tian, Y., Wang, C., Fu, Y., et al. (2018). High expression of CCR5 in melanoma enhances epithelial-mesenchymal transition and metastasis via TGFbeta1. *The Journal of Pathology*, *247*, 481–493. https://doi.org/10.1002/path.5207.
Martinez-Outschoorn, U. E., Peiris-Pages, M., Pestell, R. G., Sotgia, F., & Lisanti, M. P. (2016). Cancer metabolism: A therapeutic perspective. *Nature Reviews. Clinical Oncology*, *14*, 11–31. https://doi.org/10.1038/nrclinonc.2016.60.
Massague, J., & Obenauf, A. C. (2016). Metastatic colonization by circulating tumour cells. *Nature*, *529*(7586), 298–306. https://doi.org/10.1038/nature17038.
Moy, R. H., Huffman, A. P., Richman, L. P., Crisalli, L., Wang, X. K., Hoxie, J. A., et al. (2017). Clinical and immunologic impact of CCR5 blockade in graft-versus-host disease prophylaxis. *Blood*, *129*(7), 906–916. https://doi.org/10.1182/blood-2016-08-735076.
Nichols, W. G., Steel, H. M., Bonny, T., Adkison, K., Curtis, L., Millard, J., et al. (2008). Hepatotoxicity observed in clinical trials of aplaviroc (GW873140). *Antimicrobial Agents and Chemotherapy*, *52*(3), 858–865. https://doi.org/10.1128/AAC.00821-07.
Nishikawa, G., Kawada, K., Nakagawa, J., Toda, K., Ogawa, R., Inamoto, S., et al. (2019). Bone marrow-derived mesenchymal stem cells promote colorectal cancer progression via CCR5. *Cell Death & Disease*, *10*(4), 264. https://doi.org/10.1038/s41419-019-1508-2.
Niwa, Y., Akamatsu, H., Niwa, H., Sumi, H., Ozaki, Y., & Abe, A. (2001). Correlation of tissue and plasma RANTES levels with disease course in patients with breast or cervical cancer. *Clinical Cancer Research: An Official Journal of the American Association for Cancer Research*, 7(2), 285–289. Retrieved from http://www.ncbi.nlm.nih.gov/pubmed/11234881.
Ochoa-Callejero, L., Perez-Martinez, L., Rubio-Mediavilla, S., Oteo, J. A., Martinez, A., & Blanco, J. R. (2013). Maraviroc, a CCR5 antagonist, prevents development of hepatocellular carcinoma in a mouse model. *PLoS One*, *8*(1), e53992. https://doi.org/10.1371/journal.pone.0053992.
Olson, W. C., Rabut, G. E., Nagashima, K. A., Tran, D. N., Anselma, D. J., Monard, S. P., et al. (1999). Differential inhibition of human immunodeficiency virus type 1 fusion, gp120 binding, and CC-chemokine activity by monoclonal antibodies to CCR5. *Journal of Virology*, *73*(5), 4145–4155. Retrieved from https://www.ncbi.nlm.nih.gov/pubmed/10196311.
Perez-Martinez, L., Ochoa-Callejero, L., Rubio-Mediavilla, S., Narro, J., Bernardo, I., Oteo, J. A., et al. (2018). Maraviroc improves hepatic triglyceride content but not inflammation in a murine nonalcoholic fatty liver disease model induced by a chronic exposure to high-fat diet. *Translational Research*, *196*, 17–30. https://doi.org/10.1016/j.trsl.2018.01.004.
Pervaiz, A., Ansari, S., Berger, M. R., & Adwan, H. (2015). CCR5 blockage by maraviroc induces cytotoxic and apoptotic effects in colorectal cancer cells. *Medical Oncology*, *32*(5), 158. https://doi.org/10.1007/s12032-015-0607-x.
Qian, B. Z., & Pollard, J. W. (2010). Macrophage diversity enhances tumor progression and metastasis. *Cell*, *141*(1), 39–51. https://doi.org/10.1016/j.cell.2010.03.014.
Robinson, S. C., Scott, K. A., Wilson, J. L., Thompson, R. G., Proudfoot, A. E., & Balkwill, F. R. (2003). A chemokine receptor antagonist inhibits experimental breast tumor growth. *Cancer Research*, *63*(23), 8360–8365. Retrieved from https://www.ncbi.nlm.nih.gov/pubmed/14678997.

Ryu, H., Baek, S. W., Moon, J. Y., Jo, I. S., Kim, N., & Lee, H. J. (2018). C-C motif chemokine receptors in gastric cancer. *Molecular and Clinical Oncology*, *8*(1), 3–8. https://doi.org/10.3892/mco.2017.1470.

Samson, M., Libert, F., Doranz, B. J., Rucker, J., Liesnard, C., Farber, C. M., et al. (1996). Resistance to HIV-1 infection in caucasian individuals bearing mutant alleles of the CCR-5 chemokine receptor gene. *Nature*, *382*(6593), 722–725. https://doi.org/10.1038/382722a0.

Schlecker, E., Stojanovic, A., Eisen, C., Quack, C., Falk, C. S., Umansky, V., et al. (2012). Tumor-infiltrating monocytic myeloid-derived suppressor cells mediate CCR5-dependent recruitment of regulatory T cells favoring tumor growth. *Journal of Immunology*, *189*(12), 5602–5611. https://doi.org/10.4049/jimmunol.1201018.

Seto, T., Sam, D., & Pan, M. (2019). Mechanisms of primary and secondary resistance to immune checkpoint inhibitors in cancer. *Medical Sciences (Basel)*, 7(2), 14. https://doi.org/10.3390/medsci7020014.

Shaik, M. M., Peng, H., Lu, J., Rits-Volloch, S., Xu, C., Liao, M., et al. (2019). Structural basis of coreceptor recognition by HIV-1 envelope spike. *Nature*, *565*(7739), 318–323. https://doi.org/10.1038/s41586-018-0804-9.

Shin, N., Solomon, K., Zhou, N., Wang, K. H., Garlapati, V., Thomas, B., et al. (2011). Identification and characterization of INCB9471, an allosteric noncompetitive small-molecule antagonist of C-C chemokine receptor 5 with potent inhibitory activity against monocyte migration and HIV-1 infection. *The Journal of Pharmacology and Experimental Therapeutics*, *338*(1), 228–239. https://doi.org/10.1124/jpet.111.179531.

Sicoli, D., Jiao, X., Ju, X., Velasco-Velazquez, M., Ertel, A., Addya, S., et al. (2014). CCR5 receptor antagonists block metastasis to bone of v-Src oncogene-transformed metastatic prostate cancer cell lines. *Cancer Research*, *74*(23), 7103–7114. https://doi.org/10.1158/0008-5472.CAN-14-0612.

Sima, A. R., Sima, H. R., Rafatpanah, H., Hosseinnezhad, H., Ghaffarzadehgan, K., Valizadeh, N., et al. (2014). Serum chemokine ligand 5 (CCL5/RANTES) level might be utilized as a predictive marker of tumor behavior and disease prognosis in patients with gastric adenocarcinoma. *Journal of Gastrointestinal Cancer*, *45*(4), 476–480. https://doi.org/10.1007/s12029-014-9652-5.

Sleeman, J. P. (2015). The lymph node pre-metastatic niche. *Journal of Molecular Medicine (Berlin, Germany)*, *93*(11), 1173–1184. https://doi.org/10.1007/s00109-015-1351-6.

Soria, G., & Ben-Baruch, A. (2008). The inflammatory chemokines CCL2 and CCL5 in breast cancer. *Cancer Letters*, *267*(2), 271–285. https://doi.org/10.1016/j.canlet.2008.03.018.

Suenaga, M., Mashima, T., Kawata, N., Wakatsuki, T., Horiike, Y., Matsusaka, S., et al. (2016). Serum VEGF-A and CCL5 levels as candidate biomarkers for efficacy and toxicity of regorafenib in patients with metastatic colorectal cancer. *Oncotarget*, 7(23), 34811–34823. https://doi.org/10.18632/oncotarget.9187.

Sugasawa, H., Ichikura, T., Kinoshita, M., Ono, S., Majima, T., Tsujimoto, H., et al. (2008). Gastric cancer cells exploit CD4+ cell-derived CCL5 for their growth and prevention of CD8+ cell-involved tumor elimination. *International Journal of Cancer*, *122*(11), 2535–2541. https://doi.org/10.1002/ijc.23401.

Sugasawa, H., Ichikura, T., Tsujimoto, H., Kinoshita, M., Morita, D., Ono, S., et al. (2008). Prognostic significance of expression of CCL5/RANTES receptors in patients with gastric cancer. *Journal of Surgical Oncology*, *97*(5), 445–450. https://doi.org/10.1002/jso.20984.

Tacke, F. (2018). Cenicriviroc for the treatment of non-alcoholic steatohepatitis and liver fibrosis. *Expert Opinion on Investigational Drugs*, *27*(3), 301–311. https://doi.org/10.1080/13543784.2018.1442436.

Tan, M. C., Goedegebuure, P. S., Belt, B. A., Flaherty, B., Sankpal, N., Gillanders, W. E., et al. (2009). Disruption of CCR5-dependent homing of regulatory T cells inhibits tumor growth in a murine model of pancreatic cancer. *Journal of Immunology*, *182*(3), 1746–1755. Retrieved from https://www.ncbi.nlm.nih.gov/pubmed/19155524.

Tanabe, Y., Sasaki, S., Mukaida, N., & Baba, T. (2016). Blockade of the chemokine receptor, CCR5, reduces the growth of orthotopically injected colon cancer cells via limiting cancer-associated fibroblast accumulation. *Oncotarget*, 7(30), 48335–48345. https://doi.org/10.18632/oncotarget.10227.

Tang, Q., Jiang, J., & Liu, J. (2015). CCR5 blockade suppresses melanoma development through inhibition of IL-6-Stat3 pathway via upregulation of SOCS3. *Inflammation*, *38*(6), 2049–2056. https://doi.org/10.1007/s10753-015-0186-1.

Taube, J. M., Young, G. D., McMiller, T. L., Chen, S., Salas, J. T., Pritchard, T. S., et al. (2015). Differential expression of immune-regulatory genes associated with PD-L1 display in melanoma: Implications for PD-1 pathway blockade. *Clinical Cancer Research*, *21*(17), 3969–3976. https://doi.org/10.1158/1078-0432.CCR-15-0244.

Tsukishiro, S., Suzumori, N., Nishikawa, H., Arakawa, A., & Suzumori, K. (2006). Elevated serum RANTES levels in patients with ovarian cancer correlate with the extent of the disorder. *Gynecologic Oncology*, *102*(3), 542–545. https://doi.org/10.1016/j.ygyno.2006.01.029.

Tumeh, P. C., Harview, C. L., Yearley, J. H., Shintaku, I. P., Taylor, E. J., Robert, L., et al. (2014). PD-1 blockade induces responses by inhibiting adaptive immune resistance. *Nature*, *515*(7528), 568–571. https://doi.org/10.1038/nature13954.

Vangelista, L., & Vento, S. (2017). The expanding therapeutic perspective of CCR5 blockade. *Frontiers in Immunology*, *8*, 1981. https://doi.org/10.3389/fimmu.2017.01981.

Velasco-Velazquez, M., Jiao, X., De La Fuente, M., Pestell, T. G., Ertel, A., Lisanti, M. P., et al. (2012). CCR5 antagonist blocks metastasis of basal breast cancer cells. *Cancer Research*, *72*(15), 3839–3850. https://doi.org/10.1158/0008-5472.CAN-11-3917.

Velasco-Velazquez, M., Xolalpa, W., & Pestell, R. G. (2014). The potential to target CCL5/CCR5 in breast cancer. *Expert Opinion on Therapeutic Targets*, *18*(11), 1265–1275. https://doi.org/10.1517/14728222.2014.949238.

Wang, L. H., Lin, C. Y., Liu, S. C., Liu, G. T., Chen, Y. L., Chen, J. J., et al. (2016). CCL5 promotes VEGF-C production and induces lymphangiogenesis by suppressing miR-507 in human chondrosarcoma cells. *Oncotarget*, 7(24), 36896–36908. https://doi.org/10.18632/oncotarget.9213.

Wei, X., & Nielsen, R. (2019). CCR5-32 is deleterious in the homozygous state in humans. *Nature Medicine*, *25*(6), 909–910. https://doi.org/10.1038/s41591-019-0459-6.

Wu, Y. C., Shen, Y. C., Chang, J. W., Hsieh, J. J., Chu, Y., & Wang, C. H. (2018). Autocrine CCL5 promotes tumor progression in esophageal squamous cell carcinoma in vitro. *Cytokine*, *110*, 94–103. https://doi.org/10.1016/j.cyto.2018.04.027.

Xie, Y., Zhan, S., Ge, W., & Tang, P. (2019).?The potential risks of C-C chemokine receptor 5-edited babies in bone development. *Bone Research*, 7, 4. https://doi.org/10.1038/s41413-019-0044-0.

Yaal-Hahoshen, N., Shina, S., Leider-Trejo, L., Barnea, I., Shabtai, E. L., Azenshtein, E., et al. (2006). The chemokine CCL5 as a potential prognostic factor predicting disease progression in stage II breast cancer patients. *Clinical Cancer Research*, *12*(15), 4474–4480. https://doi.org/10.1158/1078-0432.CCR-06-0074.

Zhang, W., Xu, J., Fang, H., Tang, L., Chen, W., Sun, Q., et al. (2018). Endothelial cells promote triple-negative breast cancer cell metastasis via PAI-1 and CCL5 signaling. *The FASEB Journal*, *32*(1), 276–288. https://doi.org/10.1096/fj.201700237RR.

Zi, J., Yuan, S., Qiao, J., Zhao, K., Xu, L., Qi, K., et al. (2017). Treatment with the C-C chemokine receptor type 5 (CCR5)-inhibitor maraviroc suppresses growth and induces apoptosis of acute lymphoblastic leukemia cells. *American Journal of Cancer Research*, 7(4), 869–880. Retrieved from https://www.ncbi.nlm.nih.gov/pubmed/28469959.

CHAPTER THREE

Targeting G protein-coupled receptors in cancer therapy

Surinder M. Soond[a,*], Andrey A. Zamyatnin Jr.[a,b,*]
[a]Institute of Molecular Medicine, Sechenov First Moscow State Medical University, Moscow, Russian Federation
[b]Belozersky Institute of Physico-Chemical Biology, Lomonosov Moscow State University, Moscow, Russian Federation
*Corresponding authors: e-mail address: surinder.soond@yandex.ru; zamyat@belozersky.msu.ru

Contents

Abstract

As basic research into GPCR signaling and its association with disease has come into fruition, greater clarity has emerged with regards to how these receptors may be amenable to therapeutic intervention. As a diverse group of receptor proteins, which regulate a variety of intracellular signaling pathways, research in this area has been slow to yield tangible therapeutic agents for the treatment of a number of diseases including cancer. However, recently such research has gained momentum based on a series of studies that have sought to define GPCR proteins dynamics through the elucidation of their crystal structures. In this chapter, we define the approaches that have been adopted in developing better therapeutics directed against the specific parts of the receptor proteins, such as the extracellular and the intracellular domains, including

Advances in Cancer Research, Volume 145
ISSN 0065-230X
https://doi.org/10.1016/bs.acr.2019.11.002

the ligands and auxiliary proteins that bind them. Finally, we also briefly outline how GPCR-derived signaling transduction pathways hold great potential as additional targets.

1. Introduction

As the largest family of extracellular signaling proteins in the human genome, the G protein-coupled receptor (GPCR) family of proteins has received considerable attention due to them having emerged with greater clarity as potential therapeutic targets in many pathological conditions. Of particular relevance have been their roles in chemokine signaling, cell proliferation, invasion, survival of tumor metastases in cancer biology and the regulation of cell death (Bar-Shavit et al., 2016).

According to their protein sequence and structural similarity, GPCRs can be grouped into six main families: Class A, rhodopsin-like receptors; Class B, secretin receptor family; Class C, metabotropic glutamate/pheromone receptors; Class D, fungal mating pheromone receptors; Class E, cyclic adenosine monophosphate (cAMP) receptors; and Class F, frizzled/smoothened receptors (Audet & Bouvier, 2012; Hu, Mai, & Chen, 2017). Class A is by far the largest of the family and is therefore not surprising that it is also the most extensively studied and characterized. While originally the first GPCR described in the context of cellular transformation was identified as the MAS protein as far back as 1986 (Young, Waitches, Birchmeier, Fasano, & Wigler, 1986) and which was found to bind the angiotensin and signaled vasodilation and anti-proliferative effects (Davenport et al., 2013), such findings have indeed stimulated growth in this family of proteins. Therefrom, GPCR signaling has been linked to a number of key physiological functions, with the role of Class A receptors having been found to have significant relevance in cancer progression and which have been further sub-divided into the following four categories, namely GPCR-α, -β, -γ and -δ (Bar-Shavit et al., 2016; Luo & Yu, 2019).

Structurally, GPCRs share a common architecture and are composed of an extracellular domain (ECD), a transmembrane domain composed of seven helical structures and an intracellular domain (Lappano, Jacquot, & Maggiolini, 2018). Collectively, they can respond to a diverse number of stimuli including chemokines, light, neurotransmitters, odorants and ions (Neves, Ram, & Iyengar, 2002), which structurally alter a population of receptor proteins toward an active state (Audet & Bouvier, 2012). For example, in a ligand-receptor signaling context, the ligand occupation of the ECD induces the seven transmembrane domain (TMD) helices to undergo a conformational change (Parenti, Vigano, Newman, Milligan, & Magee, 1993; Rankovic, Brust, & Bohn, 2016; Strange, 2008), which allow the intracellular domain (ICD) to recruit a specific set of proteins that instigate intracellular signal transduction (Fig. 1). Broadly speaking, signal-inducing intracellular G-proteins (α, β or γ) are recruited to the receptor ICD, as can be signaling-suppressive GRKs, which act through ICD phosphorylation. This step also permits the recruitment of two of the four known β-arrestin proteins (Hu et al., 2013; Kovacs, Hara, Davenport, Kim, & Lefkowitz, 2009) also known as the ubiquitously expressed β-arrestins 1 and 2 (Kang, Tian, & Benovic, 2014). In addition to ligand-mediated receptor ECD stimulation, GPCRs have also been found to transduce an intracellular signal upon dimerization with other GPCRs, through post-translational (PT) modifications or in response to alterations in the pH or lipid compositions in the immediate vicinity of the GPCR environment (Dawaliby et al., 2016; Kobilka & Deupi, 2007; Ranganathan, Dror, & Carlsson, 2014; Tobin, 2008; Xue et al., 2015). Thereafter, the GPCR

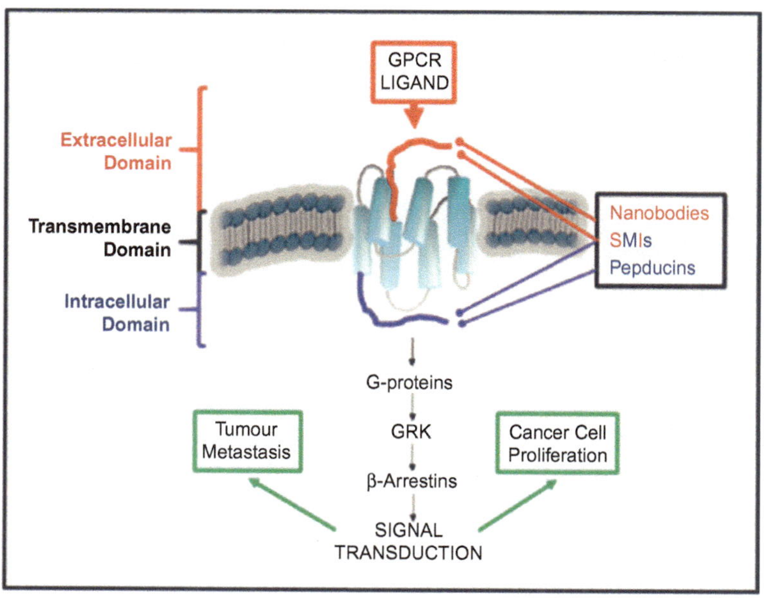

Fig. 1 Schematic detailing the GPCR structural domains and the auxiliary signaling proteins which are important as emerging therapeutic targets.

complexes are internalized to the endosomes, after which they can be recycled to the cell surface or be degraded (Bowman, Shiwarski, & Puthenveedu, 2016; Kang et al., 2014). Of importance (and from a therapeutic perspective), are the key stages in the conformational rearrangements of the ECD and the seven TMDs, upon receptor stimulation including the phosphorylation status of the ICD and its interacting G-proteins, GRKs and β-arrestins. Consequently, when taken together, the challenging picture that emerges for researchers is one that stems from the diverse intracellular signaling pathways triggered by these receptors and defining mechanistic dynamics related to receptor signaling for the rational design of high-efficacy therapeutics with enhanced specificity for disease-context.

Being mindful of the latter, more recently great progress has been made in defining the structural equilibria of some GPCRs and the findings have been encouragingly enlightening. While previously, it was generally believed that receptors were primarily in an inactive or active conformational state (Wishart, 2011), more recently some of the GPCRs have been revealed to also exist in a unique metastable 'intermediate' conformational state upon receptor occupation, which can therefore offer varying degrees of specificity for a variety of ligands (Fredriksson, Lagerstrom, Lundin, & Schioth, 2003; Jeschke, 2012; Klein Herenbrink et al., 2016). Consequently, this can also offer them the capacity to initiate different intracellular signaling cascades with varying degrees of specificity (Butcher et al., 2011) and can result from the selective activation of specific G-proteins (Rankovic et al., 2016). Collectively such findings have given rise to the growing area of 'biased agonism', which disposes pleiotropic GPCRs to initiate a specific signaling pathway with greater selectivity (Lohse et al., 2008; Violin, Crombie, Soergel, & Lark, 2014) and is an aspect of receptor signaling regulation that is intrinsic to good therapeutic design. Moreover, this has also developed the important belief that GPCRs individually may not be amenable to the exclusive intervention from just one therapeutic and that several therapeutics may hold specificity for each GPCR depending on the specific conformational state it exhibits for activating a particular intracellular signaling pathway (Bar-Shavit et al., 2016).

Another milestone in GPCR research has been the identification of the conserved ECD-specific ligand binding site, which has been identified in a growing number of GPCRs. While this has also been referred to as the 'binding pocket' it basically constitutes the orthosteric ligand binding site and is the site to which the native ligand binds and has been suggested to be very 'flexible' in nature (Dror et al., 2013; Kruse et al., 2013;

Lebon, Warne, & Tate, 2012; Wootten, Christopoulos, & Sexton, 2013). This is structurally (and exclusively) distinct to any identified therapeutic-directed allosteric binding site. While it can be difficult to discern, it does have the potential to be targeted, thus permitting the alteration of the structure and functionality of the GPCR and consequently defining the amenability of GPCR signaling for therapeutic targeting. Of equal importance are the abilities of allosterics to differentiate between receptor subtypes, thus permitting even greater selectivity in their potential targeting (Jakubik, Bacakova, El-Fakahany, & Tucek, 1997). Such allosterics have emerged to be reversible or irreversible in their ability to bind and alter receptor signaling in addition to being agonist-, antagonist- or neutral-acting in nature (Bueno et al., 2016; van der Westhuizen, Valant, Sexton, & Christopoulos, 2015). Collectively, such biophysical structural studies, when developed with chemical and computational approaches, have unveiled a wealth of information that have clearly helped in better rational design of allosteric drugs as confirmed from their association or dissociation kinetics with the GPCRs.

Such approaches in therapeutic development have also been complimented with targeting the ICD, the ICD signaling complex or the ICD-binding auxiliary proteins (such as the G-proteins and β-arrestins), thus permitting specific signaling profiles to be selectively initiated or ablated. In support of this, the recent discovery of the GPCR 'intracellular pocket' that presents itself as a 'semi-conserved' protein domain, offers wider scope for the ICD to be identified as a potential domain for rational therapeutic targeting (Dror, Dirks, Grossman, Xu, & Shaw, 2012; Park et al., 2012). Consequently, recent progress has culminated in a number of promising agonists and antagonists, through the better understanding of how GPCRs signal mainly from the combined efforts of a number of key studies and inter-disciplinary approaches founded strongly upon structural and biochemical studies.

In this chapter, we draw attention to the recent developments in therapeutic and inhibitor design for targeting GPCRs in the context of cancer progression with a particular focus on the receptor domains that have been specificity targeted by pepducin-, nanobody- and allosteric small molecule inhibitors (SMI, Ayoub, 2018; Mujic-Delic, de Wit, Verkaar, & Smit, 2014). Moreover, as an example for offering a broader context of GPCR activity, we will also touch upon how two of the main pathways emerging from GPCR signaling have been given greater consideration for therapeutic targeting as downstream signaling cascades of increasing importance in cancer progression (Liu et al., 2016; Nieto Gutierrez & McDonald, 2018; Ortiz Zacarias, Lenselink, AP, Handel, & Heitman, 2018).

2. Domain-specific GPCR-targeted therapeutics

Based on the importance of GPCRs and their role in many physiological functions and disease, research in this area has developed positively toward a translational setting over the last 10 years or so. Indeed this is reflected in the vigor and diversity with which therapeutics have been developed and which occupy the current market share of 20–50% of all therapeutics sold globally (Hauser, Attwood, Rask-Andersen, Schioth, & Gloriam, 2017; Rask-Andersen, Masuram, & Schioth, 2014; Santos et al., 2017). Generally speaking, therapeutics have broadly taken the form of agonist or antagonists and have been directed at not just the GPCRs alone, but also the receptor ligands, G-protein subunits, β-arrestins and components of their cognate downstream signaling pathways. Moreover, while SMIs have been extensively developed and exploited in this context, alternative progress has also been made in developing antibody-based therapeutics, nanobodies and peptide-based therapeutics.

Here, SMIs have demonstrated to be a popular approach but drawbacks have arisen from physiological side effects, efficacy and specificity. As an alternative, antibody therapy has been quite fruitful from the perspective of specificity and enhanced longevity but low penetrance has offered challenges due to the sheer size of the therapeutic molecule. Therefore, simultaneous new approaches through the design of 'pepducins' have offered encouraging results as an alternative. When taken with structural studies that define GPCRs as possessing 'context-dependent conformations' that can exhibit biased signaling, targeted therapeutic design has therefore not been without its challenges, but has extinguished the idea of their existing 'one therapeutic per GPCR target'.

Considering the diversity and the range of GPCRs that have been unveiled to play an important role in cellular homeostasis, this chapter will selectively approach addressing the significance of emerging therapeutics in cancer progression and the approaches being adopted. In this context, from the four groups of GPCRs, the two main groups of greatest relevance include the group A of chemokine, PAR, LPA, S1P receptors and the F group (frizzled receptors), due to its intimate connection with the Hippo/TAZ pathway for cellular growth in the latter. These will be further described based upon the regions of them that have been identified and earmarked for therapeutic intervention either through modulation of their ECD, ICD and cognate signaling pathways (Fig. 1).

3. GPCR extracellular domain-specific targeted therapy

3.1 Chemokine receptors

As the first GPCR discovered to transform cells was the MAS receptor for angiotensin signaling (Young et al., 1986), two other angiotensin receptors have also been subsequently discovered, known as AT1R and AT2R, both of which have been observed to be upregulated in expression in a number of cancer types and may therefore have good therapeutic targeting potential (Carl-McGrath, Ebert, Lendeckel, & Rocken, 2007; Liu et al., 2016). As AT1R has been seen to regulate cell proliferation and angiogenesis through VEGF, FGF and PDGF (Ager, Neo, & Christophi, 2008; Escobar, Rodriguez-Reyna, Arrieta, & Sotelo, 2004; Zhao et al., 2004), it comes as no surprise that it has therefore taken priority over AT2R in being targeted for cancer therapeutic purposes. Consequently, Candesartan (a MAS receptor antagonist) has been seen to inhibit cell proliferation in breast cancer (BC) cells by inhibiting the effects of angiotensin II and VEGF signaling (Herr et al., 2008) as a promising therapeutic. Moreover, the inhibitory actions of Losartan (Arrieta et al., 2005; Oh et al., 2016) (another MAS receptor antagonist) on BC cells demonstrated its efficacy in being able to downregulate the pro-tumourigenic effects of AT1R over-expression through XIAP, MAPK, Smad3 (or 4) (Oh et al., 2016) and NF-kappaB (Arafat et al., 2007) signaling. However, the role of AT2R in cancer does remain controversial, as groups have reported its ability to induce malignancy (Dolley-Hitze et al., 2010; Huang, Guo, Sun, Chen, & Yin, 2014) in addition to it having anti-proliferative properties (Ager, Chong, Wen, & Christophi, 2010; Du et al., 2013; Zhou et al., 2014). However, due to it inducing cardio-dysfunction, the clinical use of the ART1 antagonist has also been approached with caution (Arrieta et al., 2005).

Alternatively, other chemokine GPCRs have therefore received particular attention in their development as therapeutic targets in cancer progression. Their attractiveness has stemmed from their cytokines having the ability to enhance tumor cell mobility, homing (Zlotnik, Burkhardt, & Homey, 2011) and survival (even after their localized release, O'Hayre, Salanga, Handel, & Allen, 2008). Such chemokines can signal in an autocrine or paracrine manner (Balkwill, 2004) while also recruiting macrophages and leukocytes (which can generate a favorable microenvironment for this).

Bearing in mind the contribution made from structural studies and the identification of subdomains within the ECD of the GPCRs, the resulting potential to target chemokine GPCRs has dramatically increased, thus opening up the development of therapeutics in this area with greater possibilities of success as a major foresight. Classically, the majority of GPCR ligands bind the ECD at 'the binding pocket' or at the orthosteric binding site and the efficacy of which can be modulated by allosterics through their binding to alternative GPCR sites and thus altering the functionality of the GPCR. Consequently, the results of this permit the selective activation (or deactivation) of receptor subtypes and even allow the 'fine-tuning' of receptor signaling (Conn, Christopoulos, & Lindsley, 2009; Jakubik et al., 1997; Thal et al., 2016). For example, M1 and M2 and muscarinic receptors M1R and M2R, can be modulated through allosterics, which alter their affinity of efficacy of native ligands (Conn et al., 2009; Jakubik et al., 1997; Thal et al., 2016). Through such studies (and as proof of principle) endogenous ligands (such as $Na+$ ions) have also been seen to exhibit elements of allosteric modulation (van der Westhuizen et al., 2015). Similarly, one structural GPCR component that has gathered much attention is the 'extracellular vestibule', which occurs approximately 15A above the orthosteric binding site and is the binding site for a number of allosteric binding ligands for M2R (Dror et al., 2013).

While expression of most chemokine receptors appear to have some degree of relevance at various stages of tumor development and metastasis, in this context CXCR4 has been of great interest and has served as a reliable model with promising outcomes. CXCR4 has been seen to be over-expressed in a variety of cancers such as ovarian and ALL (Bar-Shavit et al., 2016; Chow & Luster, 2014) (Table 1). Moreover, CXCR4 is also over-expressed in BC cells (Chambers, Groom, & MacDonald, 2002; Nugent & Proia, 2017), promoting metastasis while being selectively expressed at sites of newly-established tumor metastases such as the lungs, bone marrow, liver and lymph nodes (Muller et al., 2001). It has therefore been proposed to be a good target for inducing BC inhibition (Balkwill, 2004; Burger & Kipps, 2006; Chambers et al., 2002). Consequently, approximately five major classes of CXCR4 antagonists have arisen as a result of numerous in-depth studies: (i) small modified peptides, such as the 14 amino acid peptide T140 (Tamamura et al., 2003) and its analogs, such as BKT140 (Fahham et al., 2012), POL6326 (de Nigris, Schiano, Infante, & Napoli, 2012), FC131 (Tamamura et al., 2005; Yoshikawa, Kobayashi, Oishi, Fujii, & Furuya, 2012) and TF14016 (Otani et al., 2012) which effectively inhibit CXCL12-mediated effects such as

Table 1 Chemokine GPCR-specific inhibitors and therapeutic types being used or evaluated in cancer cell models or other diseases.

GPCR (Chemokine)	Agonist/Antagonist	Target	Cancer/ *Condition*	Therapeutic type
MAS	CANDESARTAN	ATR1	Prostate	SMI
MAS	LOSARTAN	ATR1	Breast	SMI
CXCR4	T140 analogs	CXCR4	Breast, ovary, ALL	SMI
CXCR4	AMD3100 analogs	CXCR4	AML, breast	NAb
CXCR4	MSX-122	CXCR4	Breast, head/ neck, liver, melanoma	SMI
CXCR4	GSK812397	CXCR4	HIV, osteosarcoma	SMI
CXCR4	AMD11070	CXCR4	Melanoma	NAb
CXCR4	KRH-3955	CXCR4	Gastric	SMI
CXCR4	BMS936565	CXCR4	AML, CML, NHL	Ab
CXCR4	MOGAMALIZUMAB	CXCR4/ CXCL12	T-cell leukemia	Ab
CXCR4	CTCE-9908	CXCL12	Breast	Fc-CXCR4
P2Y	MRS2500	P2YR	*Vascular function*	SMI
Adenosine R	XAC	AR	*Parkinson, Asthma*	SMI
AR	CGS21680	AR ligand	*Parkinson, Asthma*	SMI
P2Y	uDPGA-B3-Ara-G4PAMA	P2YR	*Vascular function*	SMI
AR	DITC-APEC	AR/AL	*Thrombosis, inflammation*	SMI
CCR1/ CCR5	Met-RANTES	CCR1/ CCR5	Liver	SMI
CCR2	RDC018	CCL2/ CCR2	Liver	SMI

Continued

Table 1 Chemokine GPCR-specific inhibitors and therapeutic types being used or evaluated in cancer cell models or other diseases.—Cont'd

GPCR (Chemokine)	Agonist/Antagonist	Target	Cancer/ *Condition*	Therapeutic type
CCR5	MARAVIROC	CCR5/ CCL5	Liver	SMI
EP1R	AH6809	EP1R	Liver	SMI
CXCR7	CCX771	CXCL12/ CXCR4	Liver	SMI
Endothelin A R	BQ123	ET-1/ ETAR	Liver	SMI
CXCR4	AMD3100	CXCR4/ SDF1alpha	Stem cell, metastasis	SMI
SMO	GDC0448	SMO-R	Liver	SMI

Ab, antibody; ALL, acute lymphoblastic leukemia; AML, acute myeloid leukemia; HIV, human immunodeficiency virus; CML, chronic myeloid leukemia; Nab, nanobody; NHL, non-Hodgkins lymphoma; SMI, small molecule inhibitor.

migration, proliferation, and metastasis in human breast cancer MDA-MB-231 cells (Tamamura et al., 2003); (ii) small molecules, such as the CXCR4 neutral antagonist AMD3100 (also known as Plerixafor and Mozobil) (Jahnichen et al., 2010), has been derived from a collection of 15 other potential drugs (De Clercq, 2009; DiPersio, Stadtmauer, et al., 2009; DiPersio, Uy, Yasothan, & Kirkpatrick, 2009; Peled, Wald, & Burger, 2012), anti-HIV AMD070 (Nyunt et al., 2008; Stone et al., 2007), AMD11070 (O'Boyle et al., 2013), MSX-122 (Liang et al., 2012), GSK812397 (Jenkinson et al., 2010; Planesas, Perez-Nueno, Borrell, & Teixido, 2015), and KRH-3955 (Iwanaga et al., 2012; Murakami et al., 2009); (iii) antibodies to CXCR4 such as MDX-1338/BMS93656 (Kuhne et al., 2013); (iv) modified agonists and antagonists for CXCL12, such as CTCE-9908 (Drenckhan et al., 2013; Gil, Seshadri, Komorowski, Abrams, & Kozbor, 2013); and (v) microRNA approaches (Koga et al., 2014).

To compliment such efforts, several monoclonal antibodies have also arisen and are currently being evaluated at the level of clinical trials (Hutchings, Koglin, & Marshall, 2010; Klarenbeek et al., 2012; Webb, Handel, Kretz-Rommel, & Stevens, 2013), with Mogamulizumab even being approved for adult T-cell leukemia treatment (Beck & Reichert, 2012; Kaplon & Reichert, 2019). Alternatively, recent developments have given rise to a new wave of therapeutics based upon 'nanobody' technology. Whereas classically, antibody-based technology can encounter mass-production problems associated with post-translational modifications, affinity, specificity and correct assembly, such factors have been less of a consideration and are therefore less important in nanobody production. In the instance of, CXCL12/CXCR4 signaling in cancer and HIV entry (Furusato, Mohamed, Uhlen, & Rhim, 2010; Weiss, 2001), such approaches have been met with success (Jahnichen et al., 2010). Here anti-CXCR4-Nbs was successfully isolated which strongly inhibited CXCL12 induced signaling and chemotaxis alongside bi-paratopic CXCR4-Nbs which showed enhanced affinities and potencies as inverse agonists against CXCR4 (Jahnichen et al., 2010) in T-cells. No obvious drawbacks exist at this stage, except their size which is generally below the renal cut-off size of 50–60 kDa and which consequently gives them a shorter half-life in comparison to conventional full-sized antibodies in therapeutic use. They may therefore be better suited for diagnostic purposes. As an alternative (and promising approach), nanoparticle targeting in conjunction with SMIs have also been implemented for therapeutic development in cancer. Such initiatives have included developing compounds MRS2500 (de Castro et al., 2010), XAC (Kecskes, Tosh, Wei, Gao, & Jacobson, 2011), CGS21680 (Kim, Hechler, Gao, Gachet, & Jacobson, 2009), UDPGA-A3ARa-G4PAMA (de Castro et al., 2010) and DITC-APEC (Kim, Klutz, et al., 2009) with varying degrees of success (Table 1).

The collective paradigms presented by the CXCR4/CXC12 system have offered some inspiring avenues of exploration whereby other chemokine ligands have also been analogously analyzed for targeting. Such candidate ligands potentially include CXCL1–3, CXCL8, CCL2 and CCL5, which are all linked to tumor progression (Ma, Xiong, & Lee, 2018). In the context of hepatocellular carcinoma (HCC), a number of very promising therapeutics have also emerged that either modulate cellular transformation, proliferation, invasiveness and survival (Peng et al., 2018). Here, Met-RANTES (CCR1/CCR5 antagonist) decreased liver fibrosis (Barashi et al., 2013), AH6809 (EP1 antagonist) blocked proliferation (Cusimano et al., 2009), RDC018 (CCL/CCR2) inhibited HCC growth and metastasis (Li et al., 2017), AMD3100 (CXCR4 antagonist) synergized the anti-angiogenic agent Sorafenib in overcoming drug resistance in HCC (Gao et al., 2015; Liu et al., 2015), CCX771 (CXCR7 antagonist) decreased invasiveness and proliferation (Lin et al., 2014), BQ123 (endothelin type a receptor) decreased migration and invasion (Cong, Li, Shao, Li, & Yu, 2016), GDC-0449 (smoothened (Smo) receptor antagonist) decreased tumor size and cell infiltration (Jeng et al., 2015) and lastly, Maraviroc (CCR5) which offered a lower proliferation index and higher survival levels in HCC (Ochoa-Callejero et al., 2013) (Table 1).

Collectively, such approaches have given rise to a plethora of potential therapeutics that are currently being evaluated through the pipeline of development spanning their efficacy in laboratory-based models to effectiveness in the clinic and in a variety of pathological conditions. They also demonstrate that the GPCRs are amenable for therapeutic targeting using a number of different approaches. More importantly, they highlight the enthusiasm, with which the potential for anti-GPCR therapy has been embraced by the scientific community over such a short space of time and is a good example of how positive steps can be made using a multidisciplinary approach when faced with the complexities posed these receptors.

3.2 Protease-activated receptors

As the second subgroup of GPCRs of interest, the protease-activated receptors (PAR) have emerged as a very significant group of proteins in their regulation of processes tightly linked with cancer progression. Most of these attributes emerge from what has been learned about their regulation at the molecular level. More specifically, aberrant PAR1 (or PAR2) expression and signaling have been linked with lung (Jin et al., 2003; Nierodzik et al., 1998), colorectal (Darmoul, Gratio, Devaud, & Laburthe, 2004; Darmoul, Gratio, Devaud, Peiretti, & Laburthe, 2004; Darmoul, Marie, Devaud, Gratio, & Laburthe, 2001; Heider et al., 2004), prostate (Black et al., 2007; Ramsay et al., 2008), breast (Hernandez, Correa, Avila, Vela, & Perez, 2009; Su et al., 2009; Versteeg et al., 2008) and ovarian cancers (Agarwal et al., 2008; Grisaru-Granovsky et al., 2005; Jahan et al., 2007) thus identifying PAR1 as good therapeutic target (Covic & Kuliopulos, 2018). Supportingly, reduced PAR1 expression in the highly metastatic melanoma cell line A375SM attenuated *in vitro* cell invasion, proliferation and *in vivo* xenograft tumor formation and vascularization (Salah et al., 2007). Similarly, PAR1 expression was also negatively correlated with survival in patients with HER2-negative breast carcinoma (Gonda et al., 2015). Additionally, PAR4 expression is significantly increased in prostate cancer

(Black et al., 2007) and this receptor has been reported to contribute to migration of colon cancer (Gratio et al., 2010) and hepatocellular carcinoma-derived cell lines (Kaufmann et al., 2007; Kaufmann, Henklein, Henklein, & Settmacher, 2009). Clearly, such reports have highlighted solid foundations on which to develop PAR-directed therapeutics for cancer treatment. However, as PAR was originally cloned as the central protease involved in blood coagulation (Rasmussen et al., 1991; Vu, Hung, Wheaton, & Coughlin, 1991), many of the therapeutics developed against it have been seen to show diverse negative effects in cancer models and so their effectiveness has been limited to blood disorders and their applications.

PAR receptors are members of a subfamily, within the larger GPCR family of rhodopsin-like class A group receptors, which includes the four proteins PAR1-4 (Coughlin, 1994) and all of which are activated upon enzymic cleavage (Coughlin, 1999). Uniquely, PAR receptor activation occurs by the proteolytic and irreversible removal of the amino-terminal *pro*-peptide, which reveals a neoepitope termed the 'Tethered ligand' (TL; Macfarlane, Seatter, Kanke, Hunter, & Plevin, 2001; Vu et al., 1991), which can activate the receptor through intramolecular interactions. As seen for PAR1, a short peptide mimicking the revealed TL, also referred to as 'Activating peptide' (AP) can activate receptor signaling (Scarborough et al., 1992; Vu et al., 1991). While identified as a unique regulatory mechanism, it also highlighted the means by which receptor subtypes can be selectively activated and the dissection for their molecular signaling capabilities. Subsequently, APs against PAR2-4 have been identified which have presented themselves as valuable probes in furthering our understanding of these receptors under normal and disease conditions, with the potential for therapeutic development as a foresight (Macfarlane et al., 2001; Ramachandran & Hollenberg, 2008). For example, peptide antagonists that selectively bind and modulate the ECD of PAR1 include RWJ-56110 (6a) and RWJ-58259 (6b) (Andrade-Gordon et al., 2001, 1999; Maryanoff, Zhang, Andrade-Gordon, & Derian, 2003; Zhang et al., 2001) in preventative measures against thrombosis. To highlight the potential of such therapeutics, SCH530348 which is a himbacine-based antagonist of platelet PAR1, showed promising results in trials assessing safety (Clinical trial registration numbers: NCT00684203, NCT00684515, NCT00132912). Currently, it is now in phase III trials with a view to it being evaluated for efficacy in different patient populations and results from which are eagerly awaited. Similarly, E5555 is currently in phase II clinical trials in patients with acute coronary syndrome (Clinical trial registration numbers: NCT00619164 and NCT00548587), coronary artery disease (Clinical trial registration numbers: NCT00540670 and NCT00312052) (Cirino & Severino, 2010; Serebruany, Kogushi, Dastros-Pitei, Flather, & Bhatt, 2009). In addition to inhibiting PAR1, E5555 also inhibits platelet function through actions on PECAM-I, GP IIb/IIIa antigen, and activity with PAC-1, GPIb, thrombospondin, vitronectin receptor expression, and formation of platelet-monocyte aggregates (Serebruany et al., 2009). While such diverse effects may limit its use in the clinic, any benefits it may present against cancer development and progression are awaited.

Simultaneously, high-throughput screening of compound libraries for platelet PAR1 inhibitors have yielded a number of SMIs including SCH79797 (Ahn et al., 1999), FR171113 (Kato et al., 1999), F16618 (Perez et al., 2009), E5555 (Atopaxar) and SCH530348 (Vorapaxar) (Tello-Montoliu, Tomasello, Ueno, & Angiolillo, 2011) (for an overview, see Table 2). Although anti-thrombotic effects in animal models were observed for all these molecules, only Vorapaxar and Atopaxar (that bind at the TL site) have been evaluated in large-scale clinical trials. These two compounds have been noted for their high oral bioavailability, superior potency and selectivity.

In summary, while some therapeutics do appear to show some degree of potential in the context of cancer therapy, clearly the focus of immediate

Table 2 PAR GPCR-specific inhibitors and therapeutics types being used or evaluated in disease types physiological functions.

GPCR (PAR)	Agonist/ Antagonist	TARGET/Domain	*Condition/ Function*	Therapeutic type
PAR1	RWJ56110	PAR1/3/4-ECD	*Thrombosis*	Peptide
PAR1	RWJ58259	PAR1/3/4-ECD	*Thrombosis*	TL
PAR1	SCH530348	PAR1/3/4-ECD	*Atherothrombotic*	SMI
PAR1	SCH79797	PAR1/3/4-ECD	*Thrombosis*	SMI
PAR1	E555	PAR1/3/4-ECD	*Atherothrombotic*	SMI
PAR1	FR171113	PAR1/3/4-ECD	*Thrombosis*	SMI
PAR1	F16618	PAR1/3/4-ECD	*Thrombosis*	SMI
PAR1	SCH79797	PAR1/3/4-ECD	*Thrombosis*	SMI

ECD, extracellular domain; SMI, small molecule inhibitor; TL, tethered ligand.

studies appear to revolve around the regulation of thrombosis by the PAR receptors and further developments here are eagerly awaited (Table 2).

3.3 Lysophosphatidic acid receptors

As the third subgroup of interesting A-type GPCRs, the lysophosphatidic acid receptors (LPAR) are composed of six receptors (LPA1–6), some of which are expressed by endothelial cells (Aoki et al., 2002; Bian et al., 2006; Chen, Towers, & O'Connor, 2007; Contos, Ishii, & Chun, 2000; Mills & Moolenaar, 2003; Tabata, Baba, Shiraishi, Ito, & Fujita, 2007; Yang et al., 2005) and the deregulated expression of which can initiate cancer or contribute toward its progression (Mills & Moolenaar, 2003; Yang et al., 2005). For example, in breast- (Chen et al., 2007) and ovarian- (Bian et al., 2006) cancers, LPAR activates Rho-GTPases to drive cell migration and tumor formation. Consequently, some progress has been made in directing some of these receptors for therapeutic development giving rise to LPAR antagonists Kil6425 and Kil6198 and both of which have also proven their usefulness in controlling pancreatic cancer metastasis (Komachi et al., 2012; Liu et al., 2016; Yun et al., 2005). More specific to BC, the LPAR1/3 antagonist Debio0719, which is an isomer of isoxazolic competitive inhibitor Ki16425, was effectively active against LPAR1 and -3 (collectively) with a respective IC_{50} of 60 and 660 nM (Korinek et al., 1997; Valenta, Hausmann, & Basler, 2012; van Amerongen, 2012) and against the invasion of 4T1 mouse mammary cancer cells *in vitro* (David et al., 2012) or *in vivo* during the early phase of BC growth (Katanaev, Ponzielli, Semeriva, & Tomlinson, 2005) (Table 3).

Table 3 LPAR GPCR-specific inhibitors and therapeutics types being used or evaluated in cancer cell models.

GPCR (LPAR)	Agonist/ Antagonist	TARGET/ Domain	Cancer	Therapeutic type
LPA1/3	Kil6425	ECD	Pancreatic metastases	SMI
LPA1/3	Kil6198	ECD	Pancreatic	SMI
LPA1/3	Debio0719	ECD	Breast	SMI

ECD, extracellular domain; SMI, small molecule inhibitor.

3.4 Sphingosine-1-phosphate receptors

Sphingosine-1-phosphate (S1P) is a membrane-derived sphingolipid, produced from the conversion of ceramide by ceramidase sphingosine and which is then phosphorylated by the sphingokinases SK1 and SK2. These kinases are activated by GPCRs (or small G-proteins) and are strongly associated with producing S1P that is transported from the cell (Fukuhara et al., 2012) or found within intracellular compartments such as the ER (Spiegel & Milstien, 2007), nucleus (Hait et al., 2009) and the mitochondrion (Strub et al., 2011). S1P can bind five known GPCRs (called S1PR1–5) in an autocrine/paracrine manner (Alvarez, Milstien, & Spiegel, 2007; Takabe, Paugh, Milstien, & Spiegel, 2008), which are coupled to different G-proteins that signal distinct effects (Brinkmann, 2007). As S1P can modulate a variety of important pro-survival or death downstream signaling pathways, it therefore comes as little surprise that deregulated S1P signaling can have an effect on the biology of numerous cancers (Pyne & Pyne, 2010). Around 40 S1P receptors (S1PR) are know of, some of which are closely related to the LPA1-3 receptors (Im et al., 2000), which can be activated by bioactive lipids, including LPA and lysophosphatidyl serine (Kihara, Maceyka, Spiegel, & Chun, 2014) and which shows structural similarity to S1P. As each of the S1PR receptors have a distinct role in cancer progression, these receptors hold great potential in being able to be targeted in a cell context-dependent manner. For example, S1PR1 modulates neovascularization (Liu et al., 2000) and cell migration (Fisher et al., 2006; Li, Sanchez, et al., 2009; Young, Pearl, & Van Brocklyn, 2009); S1PR2 is an anti- and pro-tumor suppressor (An, Zheng, & Bleu, 2000; Cattoretti et al., 2009), migration suppressor or promoter (Lepley, Paik, Hla, & Ferrer, 2005; Patmanathan et al., 2016; Yamamura, Hakomori, Wada, & Igarashi, 2000), invasion suppressor or promoter (Arikawa et al., 2003; Young & Van Brocklyn, 2007), metastasis suppressor or promoter (Ponnusamy et al., 2012; Yamaguchi et al., 2003), neovascularization suppressor (Du et al., 2010); S1PR3 regulates pro-migration and invasiveness (Filipenko et al., 2016; Kim et al., 2011; Lee et al., 2017; Sugimoto, Takuwa, Okamoto, Sakurada, & Takuwa, 2003; Yamashita et al., 2006), neovascularization (Argraves et al., 2008; Kimura et al., 2000; Lee et al., 1999; Paik, Chae, Lee, Thangada, & Hla, 2001), drug resistance (Watson et al., 2010), stemness (Hirata et al., 2014); S1PR4 expression offers short survival and poor prognosis (Ohotski et al., 2012); and, S1PR5 has anti- and pro-tumor suppression, proliferation and migration properties (Hu et al., 2010), while also promoting autophagy (Chang et al., 2009; Huang, Chang, et al., 2014).

As S1P phospholipid is largely oncogenic, it has presented itself as a good target in therapy with the emergence of two promising therapeutics namely ABC294640 (which can inhibit SK2, cause the proteasomal degradation of SK1 and inhibit dihydroceramide desaturase; McNaughton, Pitman, Pitson, Pyne, & Pyne, 2016; Venant et al., 2015) and LT1002 (which is a monoclonal antibody; O'Brien et al., 2009), that promisingly bound S1P in picomolar quantities but was abandoned due to lack of efficacy at phase II of trials (NCT01762033). With the recent emergence of nanoparticle and pepducin technology, cell-penetrating therapeutics and S1P coated nanoparticles are also being explored to target subcellular GPCRs (Joyal, Bhosle, & Chemtob, 2015) with encouraging outcomes.

While ligand-directed therapeutics have also been in development, their cognate receptors have also been explored for their potential targeting in cancer therapy. In this instance, the promising S1P receptor antagonists FTY720 (Gilenya, Fingolimod or 2-amino-2-[2-(4-octylphenyl)ethyl] propan-1,3-diol) have been developed and which become phosphoactivated by SK2 (Paugh, Payne, Barbour, Milstien, & Spiegel, 2003). It acts as an agonist through S1P1 and S1PR3–5 through the functional antagonism of S1PR1 by prolonging its internalization and degradation (Oo et al., 2007) and has good efficacy in numerous cancer models (Filipenko et al., 2016; Kim et al., 2011; Lee et al., 2017; Yamashita et al., 2006). Mechanistically, it may also act through SK1 inhibition (White, Alshaker, Cooper, Winkler, & Pchejetski, 2016) and other targets include protein phosphatase 2A, cell cycle regulator proteins and mitochondria (White et al., 2016). While promising, it does have immunosuppressive effects due to its reducing levels of S1PR1 from the cell surface of lymphocytes, which may compromise its future development or use. Therefore, as an alternative, a novel therapeutic reagent has been formulated that makes use of a monoclonal antibody to physically sequester extracellular S1P (Sphingomab) and which has been seen to reduce tumor growth and promote anti-angiogenic effects (Visentin et al., 2006).

As the other four S1PRs have not attained the status of clinically-validated drug targets, small molecule modulators of the receptors are being developed and which do hold great potential (Kim et al., 2015; Oskeritzian et al., 2015; Yamamoto et al., 2016). For example, JTE-013 can antagonize S1PR2, S1PR4 (Long et al., 2010; Pyne & Pyne, 2011) in BC cells and S1PR3 at higher concentrations (Salomone & Waeber, 2011) *in vitro*. CYM-5478 can selectively inhibit S1P2 in neuroepithelial, ovary and oral squamous cell carcinoma cells (Herr et al., 2016; Patmanathan et al., 2016;

Table 4 S1PR GPCR-specific inhibitors and therapeutics types currently being used or evaluated in cancer cell models or cell types.

GPCR (S1PR)	Agonist/ Antagonist	TARGET/Domain	Cancer/*Cell types*	Therapeutic type
S1PR	ABC294640	SK2	Prostate	SMI
S1PR	LT1002	S1P-ligand	Ovarian carcinoma	Mab
S1PR	FTY720	S1PR3,4,5-S1P ligand	Gastric, breast, nasopharyngeal	SMI
S1PR	Sphingomab	S1P-ligand	Breast, ovarian, lung	Mab
S1P	JTE013	S1PR2,4,3-S1P ligand	Breast	SMI
S1P	CYM5478	S1PR2-S1P ligand	Ovarian, OSCC	SMI
S1P	CYM-5541	S1PR3-S1P ligand	*Cardiac myocytes*	SMI
S1P	CYM50308	S1P4-S1P ligand	–	SMI

Mab, monoclonal antibody; OSCC, oral squamous cell carcinoma; SMI, small molecule inhibitor.

Satsu et al., 2013), the agonist CYM-5541 can target S1PR3 (Jo et al., 2012) *in vitro* and the agonist CYM50308 can also act potently with good selectivity against S1PR4, with low activity toward S1PR5 (Urbano et al., 2011). However, the effects of such inhibitors in model cancer cell systems are largely yet to be validated in more detail (Table 4).

4. The intracellular domain as an attractive therapeutic target

Classically, the extracellular domains of the GPCRs have been recognized as important potential targets and are yielding some very interesting findings within the scope of cancer research and therapeutic development. However, as greater mechanistic insight into how GPCR's are regulated has emerged, a simultaneous requirement for therapeutics with greater specificity has also arisen. One region of the GPCRs that has been identified as a potential target includes the ICD, based upon its ability to selectively interact with auxiliary proteins that initiate intracellular signal transduction pathways. Generally speaking, the design of such therapeutics has been conventionally based upon identifying 'protein binding partners' with a view to disrupting their ICD-protein interacting potential and thus reducing their signaling capabilities. However, more recently, positive developments in the area of structural studies have given rise to greater rational design of drugs with some very interesting findings that have also identified new and conserved target sequences within the GPCR ICD. Consequently, the design of orthosteric ligands, allosteric inhibitors, pepducins (Carr 3rd & Benovic, 2016), intrabodies (expressed in cells to modulate G-protein or β-arrestin signaling; Shukla, 2014; Staus et al., 2014) and RNA aptamers (which can inhibit the functional coupling of β-arrestin 2 to Gαs; Kahsai et al., 2016) have been explored with very encouraging outcomes.

5. Targeting the GPCR ICD and its signaling proteins

Such developments have indeed been motivated by successful structural studies founded on the co-crystallization of GPCRs complexed with SMIs (Liu et al., 2017; Oswald et al., 2016; Zheng et al., 2016) and from studies whereby the ligands have been seen to interact with the cytoplasmic side of the GPCR, as seen with CCR2 and CCR9 (Andrews, Jones, & Wreggett, 2008; Nicholls et al., 2008; Oswald et al., 2016; Zheng et al., 2016). This has given rise to therapeutics such as Vercimon, which targets the ICD of CCR9 (Wendt & Keshav, 2015). Similarly, the interaction of CCR2 with the allosteric antagonist CCR2-RA-[92] and the orthosteric antagonist BMS-861 (Hilger, Masureel, & Kobilka, 2018) have been unveiled and of which CCR2-RA-[R] has been seen to bind the intracellular face of the GPCR. Moreover, small molecule library screening has also revealed the antagonist compound 15 (Cmpd-15) to interact with the cytoplasmic regions of the transmembrane domains for the β2 adrenergic receptor (β2AR) and to inhibit β-arrestin recruitment to the ICD (Ahn et al., 2017; Liu et al., 2017). Collectively, such observations highlight the feasibility of such approaches with the modulation of biased signaling as being an important factor that can also be a consideration (Ahn et al., 2017) in the context of β-arrestin or G protein-specific targeting (Smith, Lefkowitz, & Rajagopal, 2018; Xiao & Sun, 2018). Moreover, this has been helped through defining the existence of a conserved 'intracellular pocket' in some receptors, which constitutes a potential target sequence that overlaps the G-protein and β-arrestin binding site and which can be derived from the TM helices. For example, the glucagon receptor has an ICD allosteric site between TMs 6 and 7 and which can covalently bind cysteine residues and thus prevent outward motion of TM6 resulting in reduced receptor activation (Bueno et al., 2016). As expected, such a sequence would indeed permit the targeting of specific signal transduction pathways and may thus offer greater specificity for therapeutic targeting purposes and thus potentially permit further modulation of biased signaling (Carr 3rd & Benovic, 2016; Rasmussen et al., 2011; Shukla et al., 2014). While studies are still in their infancy, initial observations that β-arrestin 2 knockdown can reverse the effects of LPA induced effects in BC, migration and invasion indeed offer strong foundations to target arrestins in cancers (Li, Alemayehu, et al., 2009). Similarly, G-proteins can also be targeted and modulated as seen through the actions of the pertussis and cholera toxins (Gill & Meren, 1978; West Jr., Moss, Vaughan, Liu, & Liu, 1985). As an example in support of this, YM254890, which is a cyclic depsipeptide, can bind $G\alpha_q$, where the GDP bound state can be stabilized, thus inhibiting GDP/GTP exchange reactions and proteins $G\alpha_q$, $G\alpha_{11}$ and $G\alpha_{14}$ (Nishimura et al., 2010).

Peptudins are short, palmitoylated cell-penetrating peptides derived from the intracellular loop (ICL) of GPCRs and which can remain tethered to the inner leaf of the cell membrane (Covic, Gresser, Talavera, Swift, & Kuliopulos, 2002; Miller et al., 2009; O'Callaghan, Kuliopulos, & Covic, 2012; Tressel et al., 2011). Here, they can exert their effects more specifically due to ICL sequence diversity throughout the GPCR family members and directly to the GPCRs through allosteric binding site recognition (O'Callaghan et al., 2012). The power of these therapeutics lie in their ability to stabilize the GPCR in a specific conformational state, which consequently can induce a specific signaling effect. In this context, the PAR receptors have gained particular attention with some very promising outcomes. In a recent study, the PAR1 pepducin P1pal-7 (pal-KKSRALF-NH2) was reported to block MMP-1 induced PAR1 activation of the Akt survival pathway in BC cells, resulting in apoptosis of tumor xenografts and inhibition of lung metastases by up to 88% (Yang et al., 2009). While ectopic expression of functionally active PAR1 in BC cells after injection into mice, induced a hormone-refractory invasive phenotype that metastasized, cells expressing a non-signaling PAR1 did not develop invasive BC (Yang et al., 2016). In a xenograft model of BC, progression was markedly attenuated following infusion of pepducin, P1pal-7, with Taxotere compared to Taxotere alone. Metastases were also inhibited by P1pal-7 alone (Yang et al., 2009). Similarly, a tumor-targeted insertion peptide coupled to intracellular loop 3 of PAR1 also blocked BC growth (Burns & Thevenin, 2015). Additionally, P1pal-12 and P4pal-10 (targeting PAR1 and PAR4, respectively) were identified as potentially useful antagonists

(Covic et al., 2002; O'Callaghan et al., 2012) in fibroblast cells. Nevertheless, antagonists Vorapaxar, Atopaxar, and PZ-128 for PAR2 and PAR4 have undergone clinical evaluation and discouragingly do present risks of coronary heart disease (Covic & Kuliopulos, 2018). Another important pepducin includes the agonist ATI-2341 (O'Callaghan et al., 2012) for CXCR4 which acts by enhancing CXCR4 coupling with $G\alpha_i$ (to activate calcium ion flux) and by promoting a minimal interaction with $G\alpha_{13}$, GRKs and β-arrestin (Quoyer et al., 2013). Collectively, such findings highlight the importance of addressing specificity and side effects through therapeutic evaluation in the area of GPCR targeting in model systems. While delivering pepducins to tissues may present its own challenges from the perspective of penetrability (Carr 3rd & Benovic, 2016), PZ-128 (Gurbel et al., 2016) or par2-specific PZ-235 (Shearer et al., 2016) are two noteworthy exceptions as they were demonstrated to have organ- and tissue-penetrability after administration (Table 5).

Alternatively, some studies have also reported the valuable use of intrabodies (or intracellularly expressed nanobodies) (Desmyter, Spinelli, Roussel, & Cambillau, 2015) so that intracellular GPCR signaling proteins can be targeted with greater selectivity, as in the case of the β-arrestin 2 intrabody which inhibited antagonist-induced cAMP response and β2AR receptor phosphorylation (Staus et al., 2014). They were also reported to uncouple GαS and β-arrestin from the receptor as these are overlapping sites within the ICD (Komolov et al., 2017; Rasmussen et al., 2011; Shukla et al., 2014).

Table 5 GPCR ICD-specific inhibitors and therapeutics types being used or evaluated in cancer cell models and other disease conditions.

GPCR (ICD)	Agonist/ Antagonist	TARGET/ Domain	Cancer/*Condition*	Therapeutic type
CCR2	CCR2-PA	ICD	Prostate, bladder	SMI
CCR9	VERCIMON	ICD	*Bowel disease*	SMI
PAR1	P1pal-1	ICD	Breast	PEPDUCIN
PAR1/4	P1pal1-10	ICD	*Thrombosis*	PEPDUCIN
PAR1	P1pal1-12	ICD	*Thrombosis*	PEPDUCIN
PAR1	PZ-128	ICD	Breast, lung, ovarian	Intracellular Ab
PAR2	PZ-235	ICD	Liver	Intracellular Ab

Ab, antibody; SMI, small molecule inhibitor.

Similarly, β-arrestin targeting is also becoming a realistic option in cancer treatment as are targeting the GRK signaling protein kinases (Nogues et al., 2018, 2017). To date, seven serine/threonine GRKs exist which are normally phosphorylated in a ligand-dependent manner (Dhami, Babwah, Sterne-Marr, & Ferguson, 2005; Eichmann et al., 2003; Pitcher, Freedman, & Lefkowitz, 1998; Watari, Nakaya, & Kurose, 2014) resulting in receptor desensitization through β-arrestin binding (Penela, Ribas, & Mayor Jr., 2003; Sorkin & von Zastrow, 2009). Mechanistically, this happens through decreasing G-protein activation and internalization, which inhibits downstream signaling pathways related to cancer progression (Goodman Jr. et al., 1996; Ruiz-Gomez & Mayor Jr, 1997). GRK2 has been reported to be upregulated in thyroid carcinoma (Metaye, Menet, Guilhot, & Kraimps, 2002), BC (Nogues et al., 2016), while showing cell cycle suppressive effects in hepatocellular carcinoma HepG2 cells (Wei et al., 2013). GRK3 can regulate CXCR4 activation of CXCL12 and GRK3 can negatively regulated BC metastasis through this pathway (Billard et al., 2016) by enhancing migration of endothelial cells in prostate cancer progression (Li et al., 2014) and found to be expressed at high levels in oral SCC (Perez-Sayans et al., 2012). High GRK5 expression has the worse prognosis in stages II–IV glioblastoma (Kaur et al., 2013), and GRK6 expression was also correlated with hepatocellular carcinoma progression (Li, 2013). Collectively, these GRKs may be targeted through their unique kinase domains or through decreasing GRK expression through RNA aptamers (Tesmer, Lennarz, Mayer, & Tesmer, 2012). At this moment, no effective GRK inhibitors have been approved for clinical use (Homan & Tesmer, 2015) and which may possibly be due to non-specific side effects (Waldschmidt et al., 2016). However, Takeda compound 3A could inhibit GRK2 over other GRKs by 50-fold, thus potentially offering an encouraging paradigm (Thal, Yeow, Schoenau, Huber, & Tesmer, 2011) (Table 6).

Table 6 GPCR ICD signaling protein-specific inhibitors and therapeutics types being used or evaluated in cancer cell models or disease conditions.

GPCR (ICD)	Agonist/Antagonist	TARGET/Domain	Cancer/*Condition*	Therapeutic type
β2-AR	Cmpd-15	β-arrestin 2	*Thrombosis*	SMI
GPCR	YM254890	G-proteins α_q	*Thrombosis*	SMI
GPCR	TAKEDA 3A	GRK2	Breast, pancreas	APTAMER

SMI, small molecule inhibitor.

6. Targeting GPCR-specific signal transduction intermediates

The GPCR family of proteins can trigger a diverse number of signaling pathways upon receptor stimulation. While the signals generally diverge from the ICD sequestering G-proteins, GRKs and β-arrestins, the receptors can also exhibit 'crosstalk' signaling to other GPCRs thereby offering a complicated network of signaling transduction pathway activation events. While this has presented a number of challenges in the area of therapeutic design, such obstacles have not generally been seen to offer a long-lasting great degree of hindrance. With this in mind, throughout this last section, it is worth touching on a few of the signaling pathways to highlight their importance in how GPCR signals can be selectively manipulated for a desired output and the principals being employed.

As universal scaffolding and GPCR signaling transducers (Srivastava, Gupta, Gupta, & Shukla, 2015), the β-arrestins have come under the radar as potential candidates useful for therapeutic targeting the GPCRs for a specific response in downstream signaling through the use of intrabodies for targeting β-arrestins (Ghosh et al., 2017) and the use of RNA aptamers that bind them (Kotula et al., 2014). Herein, aptamers that ablated signal transduction downstream of frizzled- and smoothened-GPCR receptors were found to inhibit the growth of myelogenous leukemia cells (Kotula et al., 2014). Alternatively, some intrabodies induced the differential binding of β-arrestins for clathrin and ERK kinase, as seen with intrabody synthetic antibody fragment Fab5 in cultured embryonic kidney cells (Ghosh et al., 2017) and thus defined a novel manner in which specific signaling profiles can be activated downstream of GPCR stimulation.

7. GPCR-derived Hippo-YAP-TAZ signaling as a therapeutic target

Briefly, the Hippo signaling pathway was originally defined in *Drosophila melanogaster* during a diverse genetic screen for tumor suppressor proteins and later found to be highly conserved in mammals and central to regulation of tissue size and tumorigenesis (Johnson & Halder, 2014; Pan, 2010). The core of the pathway incorporates a number of protein kinases which converge on the phosphorylation of YAP (yes-associated protein) and TAZ (transcriptional coactivator with PDZ-binding motif) which is also known as WWTR1. Broadly speaking, inactivation of upstream kinases leads to the translocation of dephosphorylated YAP/TAZ to the nucleus, which can then interact with TEAD1–4 proteins and modulate gene expression. Conversely, upon phosphorylation of YAP/TAZ, they can be retained in the cytoplasm by the 14-3-3 proteins and regulated by ubiquitination-dependent degradation (Lei et al., 2008; Zhao et al., 2007; Zhao, Li, Tumaneng, Wang, & Guan, 2010), resulting in TEAD1-4 interacting with VGLL4 which transcriptionally represses target genes (Koontz et al., 2013; Zhang et al., 2014). Thus the transcriptional effects of the hippo pathway are quintessentially dependent on the activation of kinases upstream of YAP/TAZ.

The activation of the pathways can be triggered by a number of physiological cues through membrane-associated Neurofibromin 2 (Striedinger et al., 2008), cell-cell contact (Zhao et al., 2007) mechanical forces imposed by the ECM, cell geometry and small GTPases, such as RAP2 (Dupont et al., 2011; Meng et al., 2018), hypoxia, oxidative and ER stress (Hong et al., 2017; Lin et al., 2017; Ma et al., 2015, 2016; Shao et al., 2014; Wu et al., 2015) and the GPCR signaling pathways in cancer (Miller et al., 2012; Yu, Mo, & Guan, 2012; Yu, Zhao, et al., 2012). In the latter context, YAP/TAZ can behave as oncoproteins, whereas the upstream kinases can take on the role of tumor suppressors. Hippo signaling can induce cell proliferation (Hong et al., 1893; Lei et al., 2008), transformation through EMT (Chan et al., 2008; Cordenonsi et al., 2011; Overholtzer et al., 2006), cancer (Cordenonsi et al., 2011) and stemness in stem cells (Dawood, Austin, & Cristofanilli, 2014). For example, GPER, LPA and S1P receptors are the most characterized for Hippo activation, while other activators include protons (Zhu et al., 2015), purines and adenosine (Mo, Yu, Gong, Brown, & Guan, 2012; Wang et al., 2017; Wennmann et al., 2014; Yu, Miyamoto, & Brown, 2016). Additional GPCRs that have been strongly linked with YAP/TAZ signaling include the PARs (such as ETAR, EP2/4), frizzled and CXCR4.

The characterization of the GPCRs in TAZ/Hippo regulation has given rise to the development of a number of key therapeutics because of an increase in the number of protein intermediates that transduce YAP/TAZ signals, especially with the help of cancer genome sequencing which has revealed a number of interesting mutations within signaling intermediates in multiple tumor types (Nieto Gutierrez & McDonald, 2018). Consequently, targeting TEAD (Bum-Erdene et al., 2019; Chan et al., 2016; Li et al., 2018; Noland et al., 2016), Gαs (Bao et al., 2011;

Table 7 GPCR-mediated TAZ/Hippo-specific inhibitors and therapeutics types being used or evaluated in cancer cell models or cell lines.

GPCR pathway	Agonist/Antagonist	TARGET/Domain	Cancer/*Cell type*	Therapeutic type
Frizzled/Smo	β-arr2A3	β-arrestin	CML	Aptamer
Frizzled/Smo	FAB5	β-arrestin	*HEK293*	Intrabody
Frizzled/ YAP-TAZ	FR900359	LPA,S1P, G-pro,TEAD	Uveal melanoma	SMI
Frizzled/Smo	125OSTAT101	FRD10	Synovial sarcoma	Mab
FD1-, 2-, 5-, 8-R	VANTICTUMAB	ECD	Breast, lung, PDAC	Mab
Hippo	STATINS	YAP/TAZ PTM	Breast	SMI
Wnt	CGX1321	WNT ligand	Pancreatic	SMI
Wnt	IPAFRICEPT	WNT ligand	Pancreatic, liver	FZD8-Fc fusion

CML, chronic myeloid leukemia; Mab, monoclonal antibody; SMI, small molecule inhibitor.

Dethlefsen et al., 2017), LPA and S1P (Nieto Gutierrez & McDonald, 2018; Park & Guan, 2013), $G\alpha_{q/11}$, MAPK/YAP (by FR900359) (Annala et al., 2019; Onken et al., 2018), PKA (by Rolipram), HMG-CoA inhibition (by statins) (Sorrentino et al., 2014) and YAP/TEAD (Stanger, 2012) have all been shown encouraging results in targeting Hippo signaling and cancer ablation (Table 7).

8. GPCR-derived Wnt signaling as a therapeutic target

Other signaling transduction pathways that converge on the TEAD transcription factor with significant importance in cancer progression are the Wnt-ligand and frizzled signaling pathways. Briefly, Wnt proteins were identified initially in Drosophila and are involved in normal tissue development and homeostasis. The GPCR frizzled receptor signal antagonizes the destruction of β-catenin (in the absence of which, β-catenin is continuously degraded). Conversely, stabilized β-catenin can translocate to the nucleus where it can modulate the activity of the LEF/TCF transcription factors (Korinek et al., 1997; Nusse, 2005). The stabilized form of β-catenin has been found in a number of cancers (Nusse & Varmus, 1982) especially colon carcinoma (Nishisho et al., 1991), due to the occurrence of oncogenic mutations within its amino-terminal phosphorylation site (van Amerongen, 2012).

A direct link between human cancer and Wnt signaling was established first in 1991 when mutations of the adenomatous polyposis coli (APC) gene were uncovered as the underlying cause of familial adenomatous polyposis (Nishisho et al., 1991). To date, a number of potential therapeutics have been developed against the ligand and/or receptor for Wnt. For example, CGX1321 (Pai et al., 2017), Ipafricept (Le, McDermott, & Jimeno, 2015) are Wnt inhibitors and their efficacy in the clinic is currently under evaluation and so far shows good promise (Weekes, Berlin, Lenz, et al., 2016). Another phase Ib trial of Ipafricept in HCC in association with Sorafenib is also currently underway (NCT02069145). Similarly, antagonists for the FZD receptors have also emerged such as the monoclonal antibody 125OSTA101 which blocks FRD10 (Giraudet, Badel, Cassier, et al., 2014), Vantictumab (which can bind FZD1, 2, 5, 8) and is currently in trials with BC, non-small cell lung adenocarcinoma and pancreatic ductal adenocarcinoma (PDAC) (Smith, Rosen, Chugh, et al., 2013) (Table 7).

In summary, a number of approaches are being undertaken to target GPCR-specific Wnt and FZD/Smo pathways, using a variety of approaches, either in combined treatment regiments or as stand-alone therapies. While such studies are in their infancy, they are based on sound scientific approaches that will no doubt yield some interesting outcomes in the areas of targeting GPCR in cancer.

9. Future perspectives

As highlighted herein, the size and scope of GPCR translational research in cancer has indeed gained greater importance over the recent years, particularly as 108 GPCRs are the targets of 475 FDA-approved drugs, which account for over 108 billion annual USD in sales. While initial studies have focused on defining with greater clarity the signaling pathways that originate from this very diverse family of receptors and ligands, these areas have sufficiently developed mechanistically to permit the rational and targeted design of therapeutics due to the flourishing research in the key areas of structural studies. Consequently, a number of key areas of research have arisen that are focused on targeting the GPCR-ligand, -ECD, -ICD and -signaling proteins and all of which are not mutually exclusive in striving toward improving the selectivity with which therapeutics can act. When taken with the various types of therapeutics that have been developed and the approaches that have been adopted in relation to the various receptor or signaling protein targets involved in GPCR, the potential for success seems highly likely. A direct reflection of this is embodied in the number of therapeutics that are currently being trialed clinically in relation to 5 years ago and which now must move into a cancer context. From this perspective, the future for GPCR-directed therapeutic intervention looks very promising.

Funding

This research was supported by the Russian Science Foundation (grant # 16-15-10410).

Conflicts of interest

The authors declare no conflict of interest.

References

Agarwal, A., Covic, L., Sevigny, L. M., Kaneider, N. C., Lazarides, K., Azabdaftari, G., et al. (2008). Targeting a metalloprotease-PAR1 signaling system with cell-penetrating pepducins inhibits angiogenesis, ascites, and progression of ovarian cancer. *Molecular Cancer Therapeutics*, *7*, 2746–2757. https://doi.org/10.1158/1535-7163.MCT-08-0177.

Ager, E. I., Chong, W. W., Wen, S. W., & Christophi, C. (2010). Targeting the angiotensin II type 2 receptor (AT2R) in colorectal liver metastases. *Cancer Cell International*, *10*, 19. https://doi.org/10.1186/1475-2867-10-19.

Ager, E. I., Neo, J., & Christophi, C. (2008). The renin-angiotensin system and malignancy. *Carcinogenesis*, *29*, 1675–1684. https://doi.org/10.1093/carcin/bgn171.

Ahn, H. S., Arik, L., Boykow, G., Burnett, D. A., Caplen, M. A., Czarniecki, M., et al. (1999). Structure-activity relationships of pyrroloquinazolines as thrombin receptor antagonists. *Bioorganic & Medicinal Chemistry Letters*, *9*, 2073–2078. https://doi.org/10.1016/s0960-894x(99)00339-x.

Ahn, S., Kahsai, A. W., Pani, B., Wang, Q. T., Zhao, S., Wall, A. L., et al. (2017). Allosteric "beta-blocker" isolated from a DNA-encoded small molecule library. *Proceedings of the National Academy of Sciences of the United States of America*, *114*, 1708–1713. https://doi.org/10.1073/pnas.1620645114.

Alvarez, S. E., Milstien, S., & Spiegel, S. (2007). Autocrine and paracrine roles of sphingosine-1-phosphate. *Trends in Endocrinology and Metabolism*, *18*, 300–307. https://doi.org/10.1016/j.tem.2007.07.005.

An, S., Zheng, Y., & Bleu, T. (2000). Sphingosine 1-phosphate-induced cell proliferation, survival, and related signaling events mediated by G protein-coupled receptors Edg3 and Edg5. *The Journal of Biological Chemistry*, *275*, 288–296. https://doi.org/10.1074/jbc.275.1.288.

Andrade-Gordon, P., Derian, C. K., Maryanoff, B. E., Zhang, H. C., Addo, M. F., Cheung, W., et al. (2001). Administration of a potent antagonist of protease-activated receptor-1 (PAR-1) attenuates vascular restenosis following balloon angioplasty in rats. *The Journal of Pharmacology and Experimental Therapeutics*, *298*, 34–42.

Andrade-Gordon, P., Maryanoff, B. E., Derian, C. K., Zhang, H. C., Addo, M. F., Darrow, A. L., et al. (1999). Design, synthesis, and biological characterization of a peptide-mimetic antagonist for a tethered-ligand receptor. *Proceedings of the National Academy of Sciences of the United States of America*, *96*, 12257–12262. https://doi.org/10.1073/pnas.96.22.12257.

Andrews, G., Jones, C., & Wreggett, K. A. (2008). An intracellular allosteric site for a specific class of antagonists of the CC chemokine G protein-coupled receptors CCR4 and CCR5. *Molecular Pharmacology, 73*, 855–867. https://doi.org/10.1124/mol.107.039321.

Annala, S., Feng, X., Shridhar, N., Eryilmaz, F., Patt, J., Yang, J., et al. (2019). Direct targeting of Galphaq and Galpha11 oncoproteins in cancer cells. *Science Signaling, 12*, eaau5948. https://doi.org/10.1126/scisignal.aau5948.

Aoki, J., Taira, A., Takanezawa, Y., Kishi, Y., Hama, K., Kishimoto, T., et al. (2002). Serum lysophosphatidic acid is produced through diverse phospholipase pathways. *The Journal of Biological Chemistry, 277*, 48737–48744. https://doi.org/10.1074/jbc.M206812200.

Arafat, H. A., Gong, Q., Chipitsyna, G., Rizvi, A., Saa, C. T., & Yeo, C. J. (2007). Antihypertensives as novel antineoplastics: Angiotensin-I-converting enzyme inhibitors and angiotensin II type 1 receptor blockers in pancreatic ductal adenocarcinoma. *Journal of the American College of Surgeons, 204*, 996–1005. discussion 1005–1006. https://doi.org/10.1016/j.jamcollsurg.2007.01.067.

Argraves, K. M., Gazzolo, P. J., Groh, E. M., Wilkerson, B. A., Matsuura, B. S., Twal, W. O., et al. (2008). High density lipoprotein-associated sphingosine 1-phosphate promotes endothelial barrier function. *The Journal of Biological Chemistry, 283*, 25074–25081. https://doi.org/10.1074/jbc.M801214200.

Arikawa, K., Takuwa, N., Yamaguchi, H., Sugimoto, N., Kitayama, J., Nagawa, H., et al. (2003). Ligand-dependent inhibition of B16 melanoma cell migration and invasion via endogenous S1P2 G protein-coupled receptor. Requirement of inhibition of cellular RAC activity. *The Journal of Biological Chemistry, 278*, 32841–32851. https://doi.org/10.1074/jbc.M305024200.

Arrieta, O., Guevara, P., Escobar, E., Garcia-Navarrete, R., Pineda, B., & Sotelo, J. (2005). Blockage of angiotensin II type I receptor decreases the synthesis of growth factors and induces apoptosis in C6 cultured cells and C6 rat glioma. *British Journal of Cancer, 92*, 1247–1252. https://doi.org/10.1038/sj.bjc.6602483.

Audet, M., & Bouvier, M. (2012). Restructuring G-protein-coupled receptor activation. *Cell, 151*, 14–23. https://doi.org/10.1016/j.cell.2012.09.003.

Ayoub, M. A. (2018). Small molecules targeting heterotrimeric G proteins. *European Journal of Pharmacology, 826*, 169–178. https://doi.org/10.1016/j.ejphar.2018.03.003.

Balkwill, F. (2004). Cancer and the chemokine network. *Nature Reviews. Cancer, 4*, 540–550. https://doi.org/10.1038/nrc1388.

Bao, Y., Nakagawa, K., Yang, Z., Ikeda, M., Withanage, K., Ishigami-Yuasa, M., et al. (2011). A cell-based assay to screen stimulators of the Hippo pathway reveals the inhibitory effect of dobutamine on the YAP-dependent gene transcription. *Journal of Biochemistry, 150*, 199–208. https://doi.org/10.1093/jb/mvr063.

Barashi, N., Weiss, I. D., Wald, O., Wald, H., Beider, K., Abraham, M., et al. (2013). Inflammation-induced hepatocellular carcinoma is dependent on CCR5 in mice. *Hepatology, 58*, 1021–1030. https://doi.org/10.1002/hep.26403.

Bar-Shavit, R., Maoz, M., Kancharla, A., Nag, J. K., Agranovich, D., Grisaru-Granovsky, S., et al. (2016). G protein-coupled receptors in cancer. *International Journal of Molecular Sciences, 17*, 1320. https://doi.org/10.3390/ijms17081320.

Beck, A., & Reichert, J. M. (2012). Marketing approval of mogamulizumab: A triumph for glyco-engineering. *mAbs, 4*, 419–425. https://doi.org/10.4161/mabs.20996.

Bian, D., Mahanivong, C., Yu, J., Frisch, S. M., Pan, Z. K., Ye, R. D., et al. (2006). The G12/13-RhoA signaling pathway contributes to efficient lysophosphatidic acid-stimulated cell migration. *Oncogene, 25*, 2234–2244. https://doi.org/10.1038/sj.onc.1209261.

Billard, M. J., Fitzhugh, D. J., Parker, J. S., Brozowski, J. M., McGinnis, M. W., Timoshchenko, R. G., et al. (2016). G protein coupled receptor kinase 3 regulates breast cancer migration, invasion, and metastasis. *PLoS One, 11,* e0152856. https://doi.org/10.1371/journal.pone.0152856.

Black, P. C., Mize, G. J., Karlin, P., Greenberg, D. L., Hawley, S. J., True, L. D., et al. (2007). Overexpression of protease-activated receptors-1,-2, and-4 (PAR-1, -2, and -4) in prostate cancer. *Prostate, 67,* 743–756. https://doi.org/10.1002/pros.20503.

Bowman, S. L., Shiwarski, D. J., & Puthenveedu, M. A. (2016). Distinct G protein-coupled receptor recycling pathways allow spatial control of downstream G protein signaling. *The Journal of Cell Biology, 214,* 797–806. https://doi.org/10.1083/jcb.201512068.

Brinkmann, V. (2007). Sphingosine 1-phosphate receptors in health and disease: Mechanistic insights from gene deletion studies and reverse pharmacology. *Pharmacology & Therapeutics, 115,* 84–105. https://doi.org/10.1016/j.pharmthera.2007.04.006.

Bueno, A. B., Showalter, A. D., Wainscott, D. B., Stutsman, C., Marin, A., Ficorilli, J., et al. (2016). Positive allosteric modulation of the glucagon-like peptide-1 receptor by diverse electrophiles. *The Journal of Biological Chemistry, 291,* 10700–10715. https://doi.org/10.1074/jbc.M115.696039.

Bum-Erdene, K., Zhou, D., Gonzalez-Gutierrez, G., Ghozayel, M. K., Si, Y., Xu, D., et al. (2019). Small-molecule covalent modification of conserved cysteine leads to allosteric inhibition of the TEADYap protein-protein interaction. *Cell Chemical Biology, 26,* 378–389.e313. https://doi.org/10.1016/j.chembiol.2018.11.010.

Burger, J. A., & Kipps, T. J. (2006). CXCR4: A key receptor in the crosstalk between tumor cells and their microenvironment. *Blood, 107,* 1761–1767. https://doi.org/10.1182/blood-2005-08-3182.

Burns, K. E., & Thevenin, D. (2015). Down-regulation of PAR1 activity with a pHLIP-based allosteric antagonist induces cancer cell death. *The Biochemical Journal, 472,* 287–295. https://doi.org/10.1042/BJ20150876.

Butcher, A. J., Prihandoko, R., Kong, K. C., McWilliams, P., Edwards, J. M., Bottrill, A., et al. (2011). Differential G-protein-coupled receptor phosphorylation provides evidence for a signaling bar code. *The Journal of Biological Chemistry, 286,* 11506–11518. https://doi.org/10.1074/jbc.M110.154526.

Carl-McGrath, S., Ebert, M. P., Lendeckel, U., & Rocken, C. (2007). Expression of the local angiotensin II system in gastric cancer may facilitate lymphatic invasion and nodal spread. *Cancer Biology & Therapy, 6,* 1218–1226. https://doi.org/10.4161/cbt.6.8.4412.

Carr, R., 3rd, & Benovic, J. L. (2016). From biased signalling to polypharmacology: Unlocking unique intracellular signalling using pepducins. *Biochemical Society Transactions, 44,* 555–561. https://doi.org/10.1042/BST20150230.

Cattoretti, G., Mandelbaum, J., Lee, N., Chaves, A. H., Mahler, A. M., Chadburn, A., et al. (2009). Targeted disruption of the S1P2 sphingosine 1-phosphate receptor gene leads to diffuse large B-cell lymphoma formation. *Cancer Research, 69,* 8686–8692. https://doi.org/10.1158/0008-5472.CAN-09-1110.

Chambers, A. F., Groom, A. C., & MacDonald, I. C. (2002). Dissemination and growth of cancer cells in metastatic sites. *Nature Reviews. Cancer, 2,* 563–572. https://doi.org/10.1038/nrc865.

Chan, P., Han, X., Zheng, B., DeRan, M., Yu, J., Jarugumilli, G. K., et al. (2016). Autopalmitoylation of TEAD proteins regulates transcriptional output of the Hippo pathway. *Nature Chemical Biology, 12,* 282–289. https://doi.org/10.1038/nchembio.2036.

Chan, S. W., Lim, C. J., Guo, K., Ng, C. P., Lee, I., Hunziker, W., et al. (2008). A role for TAZ in migration, invasion, and tumorigenesis of breast cancer cells. *Cancer Research, 68,* 2592–2598. https://doi.org/10.1158/0008-5472.CAN-07-2696.

Chang, C. L., Ho, M. C., Lee, P. H., Hsu, C. Y., Huang, W. P., & Lee, H. (2009). S1P(5) is required for sphingosine 1-phosphate-induced autophagy in human prostate cancer

PC-3 cells. *American Journal of Physiology Cell Physiology*, *297*, C451–C458. https://doi.org/10.1152/ajpcell.00586.2008.

Chen, M., Towers, L. N., & O'Connor, K. L. (2007). LPA2 (EDG4) mediates Rho-dependent chemotaxis with lower efficacy than LPA1 (EDG2) in breast carcinoma cells. *American Journal of Physiology. Cell Physiology*, *292*, C1927–C1933. https://doi.org/10.1152/ajpcell.00400.2006.

Chow, M. T., & Luster, A. D. (2014). Chemokines in cancer. *Cancer Immunology Research*, *2*, 1125–1131. https://doi.org/10.1158/2326-6066.CIR-14-0160.

Cirino, G., & Severino, B. (2010). Thrombin receptors and their antagonists: An update on the patent literature. *Expert Opinion on Therapeutic Patents*, *20*, 875–884. https://doi.org/10.1517/13543776.2010.487864.

Cong, N., Li, Z., Shao, W., Li, J., & Yu, S. (2016). Activation of ETA receptor by endothelin-1 induces hepatocellular carcinoma cell migration and invasion via ERK1/2 and AKT signaling pathways. *The Journal of Membrane Biology*, *249*, 119–128. https://doi.org/10.1007/s00232-015-9854-1.

Conn, P. J., Christopoulos, A., & Lindsley, C. W. (2009). Allosteric modulators of GPCRs: A novel approach for the treatment of CNS disorders. *Nature Reviews. Drug Discovery*, *8*, 41–54. https://doi.org/10.1038/nrd2760.

Contos, J. J., Ishii, I., & Chun, J. (2000). Lysophosphatidic acid receptors. *Molecular Pharmacology*, *58*, 1188–1196. https://doi.org/10.1124/mol.58.6.1188.

Cordenonsi, M., Zanconato, F., Azzolin, L., Forcato, M., Rosato, A., Frasson, C., et al. (2011). The Hippo transducer TAZ confers cancer stem cell-related traits on breast cancer cells. *Cell*, *147*, 759–772. https://doi.org/10.1016/j.cell.2011.09.048.

Coughlin, S. R. (1994). Protease-activated receptors start a family. *Proceedings of the National Academy of Sciences of the United States of America*, *91*, 9200–9202. https://doi.org/10.1073/pnas.91.20.9200.

Coughlin, S. R. (1999). How the protease thrombin talks to cells. *Proceedings of the National Academy of Sciences of the United States of America*, *96*, 11023–11027. https://doi.org/10.1073/pnas.96.20.11023.

Covic, L., Gresser, A. L., Talavera, J., Swift, S., & Kuliopulos, A. (2002). Activation and inhibition of G protein-coupled receptors by cell-penetrating membrane-tethered peptides. *Proceedings of the National Academy of Sciences of the United States of America*, *99*, 643–648. https://doi.org/10.1073/pnas.022460899.

Covic, L., & Kuliopulos, A. (2018). Protease-activated receptor 1 as therapeutic target in breast, lung, and ovarian cancer: Pepducin approach. *International Journal of Molecular Sciences*, *19*, 2237. https://doi.org/10.3390/ijms19082237.

Cusimano, A., Fodera, D., Lampiasi, N., Azzolina, A., Notarbartolo, M., Giannitrapani, L., et al. (2009). Prostaglandin E2 receptors and COX enzymes in human hepatocellular carcinoma: Role in the regulation of cell growth. *Annals of the New York Academy of Sciences*, *1155*, 300–308. https://doi.org/10.1111/j.1749-6632.2009.03701.x.

Darmoul, D., Gratio, V., Devaud, H., & Laburthe, M. (2004). Protease-activated receptor 2 in colon cancer: trypsin-induced MAPK phosphorylation and cell proliferation are mediated by epidermal growth factor receptor transactivation. *The Journal of Biological Chemistry*, *279*, 20927–20934. https://doi.org/10.1074/jbc.M401430200.

Darmoul, D., Gratio, V., Devaud, H., Peiretti, F., & Laburthe, M. (2004). Activation of proteinase-activated receptor 1 promotes human colon cancer cell proliferation through epidermal growth factor receptor transactivation. *Molecular Cancer Research*, *2*, 514–522.

Darmoul, D., Marie, J. C., Devaud, H., Gratio, V., & Laburthe, M. (2001). Initiation of human colon cancer cell proliferation by trypsin acting at protease-activated receptor-2. *British Journal of Cancer*, *85*, 772–779. https://doi.org/10.1054/bjoc.2001.1976.

Davenport, A. P., Alexander, S. P., Sharman, J. L., Pawson, A. J., Benson, H. E., Monaghan, A. E., et al. (2013). International union of basic and clinical pharmacology. LXXXVIII. G protein-coupled receptor list: Recommendations for new pairings with cognate ligands. *Pharmacological Reviews*, *65*, 967–986. https://doi.org/10.1124/pr.112.007179.

David, M., Ribeiro, J., Descotes, F., Serre, C. M., Barbier, M., Murone, M., et al. (2012). Targeting lysophosphatidic acid receptor type 1 with Debio 0719 inhibits spontaneous metastasis dissemination of breast cancer cells independently of cell proliferation and angiogenesis. *International Journal of Oncology*, *40*, 1133–1141. https://doi.org/10.3892/ijo.2011.1309.

Dawaliby, R., Trubbia, C., Delporte, C., Masureel, M., Van Antwerpen, P., Kobilka, B. K., et al. (2016). Allosteric regulation of G protein-coupled receptor activity by phospholipids. *Nature Chemical Biology*, *12*, 35–39. https://doi.org/10.1038/nchembio.1960.

Dawood, S., Austin, L., & Cristofanilli, M. (2014). Cancer stem cells: Implications for cancer therapy. *Oncology (Williston Park)*, *28*, 1101–1107. 1110.

de Castro, S., Maruoka, H., Hong, K., Kilbey, S. M., 2nd, Costanzi, S., Hechler, B., et al. (2010). Functionalized congeners of P2Y1 receptor antagonists: 2-Alkynyl (N)-methanocarba 2'-deoxyadenosine 3',5'-bisphosphate analogues and conjugation to a polyamidoamine (PAMAM) dendrimer carrier. *Bioconjugate Chemistry*, *21*, 1190–1205. https://doi.org/10.1021/bc900569u.

De Clercq, E. (2009). The AMD3100 story: The path to the discovery of a stem cell mobilizer (Mozobil). *Biochemical Pharmacology*, *77*, 1655–1664. https://doi.org/10.1016/j.bcp.2008.12.014.

de Nigris, F., Schiano, C., Infante, T., & Napoli, C. (2012). CXCR4 inhibitors: Tumor vasculature and therapeutic challenges. *Recent Patents on Anti-Cancer Drug Discovery*, *7*, 251–264.

Desmyter, A., Spinelli, S., Roussel, A., & Cambillau, C. (2015). Camelid nanobodies: Killing two birds with one stone. *Current Opinion in Structural Biology*, *32*, 1–8. https://doi.org/10.1016/j.sbi.2015.01.001.

Dethlefsen, C., Hansen, L. S., Lillelund, C., Andersen, C., Gehl, J., Christensen, J. F., et al. (2017). Exercise-induced catecholamines activate the Hippo tumor suppressor pathway to reduce risks of breast cancer development. *Cancer Research*, *77*, 4894–4904. https://doi.org/10.1158/0008-5472.CAN-16-3125.

Dhami, G. K., Babwah, A. V., Sterne-Marr, R., & Ferguson, S. S. (2005). Phosphorylation-independent regulation of metabotropic glutamate receptor 1 signaling requires g protein-coupled receptor kinase 2 binding to the second intracellular loop. *The Journal of Biological Chemistry*, *280*, 24420–24427. https://doi.org/10.1074/jbc.M501650200.

DiPersio, J. F., Stadtmauer, E. A., Nademanee, A., Micallef, I. N., Stiff, P. J., Kaufman, J. L., et al. (2009). Plerixafor and G-CSF versus placebo and G-CSF to mobilize hematopoietic stem cells for autologous stem cell transplantation in patients with multiple myeloma. *Blood*, *113*, 5720–5726. https://doi.org/10.1182/blood-2008-08-174946.

DiPersio, J. F., Uy, G. L., Yasothan, U., & Kirkpatrick, P. (2009). Plerixafor. *Nature Reviews. Drug Discovery*, *8*, 105–106. https://doi.org/10.1038/nrd2819.

Dolley-Hitze, T., Jouan, F., Martin, B., Mottier, S., Edeline, J., Moranne, O., et al. (2010). Angiotensin-2 receptors (AT1-R and AT2-R), new prognostic factors for renal clear-cell carcinoma? *British Journal of Cancer*, *103*, 1698–1705. https://doi.org/10.1038/sj.bjc.6605866.

Drenckhan, A., Kurschat, N., Dohrmann, T., Raabe, N., Koenig, A. M., Reichelt, U., et al. (2013). Effective inhibition of metastases and primary tumor growth with CTCE-9908 in esophageal cancer. *The Journal of Surgical Research*, *182*, 250–256. https://doi.org/10.1016/j.jss.2012.09.035.

Dror, R. O., Dirks, R. M., Grossman, J. P., Xu, H., & Shaw, D. E. (2012). Biomolecular simulation: A computational microscope for molecular biology. *Annual Review of Biophysics*, *41*, 429–452. https://doi.org/10.1146/annurev-biophys-042910-155245.

Dror, R. O., Green, H. F., Valant, C., Borhani, D. W., Valcourt, J. R., Pan, A. C., et al. (2013). Structural basis for modulation of a G-protein-coupled receptor by allosteric drugs. *Nature*, *503*, 295–299. https://doi.org/10.1038/nature12595.

Du, H., Liang, Z., Zhang, Y., Jie, F., Li, J., Fei, Y., et al. (2013). Effects of angiotensin II type 2 receptor overexpression on the growth of hepatocellular carcinoma cells in vitro and in vivo. *PLoS One*, *8*, e83754. https://doi.org/10.1371/journal.pone.0083754.

Du, W., Takuwa, N., Yoshioka, K., Okamoto, Y., Gonda, K., Sugihara, K., et al. (2010). S1P(2), the G protein-coupled receptor for sphingosine-1-phosphate, negatively regulates tumor angiogenesis and tumor growth in vivo in mice. *Cancer Research*, *70*, 772–781. https://doi.org/10.1158/0008-5472.CAN-09-2722.

Dupont, S., Morsut, L., Aragona, M., Enzo, E., Giulitti, S., Cordenonsi, M., et al. (2011). Role of YAP/TAZ in mechanotransduction. *Nature*, *474*, 179–183. https://doi.org/10.1038/nature10137.

Eichmann, T., Lorenz, K., Hoffmann, M., Brockmann, J., Krasel, C., Lohse, M. J., et al. (2003). The amino-terminal domain of G-protein-coupled receptor kinase 2 is a regulatory Gbeta gamma binding site. *The Journal of Biological Chemistry*, *278*, 8052–8057. https://doi.org/10.1074/jbc.M204795200.

Escobar, E., Rodriguez-Reyna, T. S., Arrieta, O., & Sotelo, J. (2004). Angiotensin II, cell proliferation and angiogenesis regulator: Biologic and therapeutic implications in cancer. *Current Vascular Pharmacology*, *2*, 385–399.

Fahham, D., Weiss, I. D., Abraham, M., Beider, K., Hanna, W., Shlomai, Z., et al. (2012). In vitro and in vivo therapeutic efficacy of CXCR4 antagonist BKT140 against human non-small cell lung cancer. *The Journal of Thoracic and Cardiovascular Surgery*, *144*, 1167–1175.e1161. https://doi.org/10.1016/j.jtcvs.2012.07.031.

Filipenko, I., Schwalm, S., Reali, L., Pfeilschifter, J., Fabbro, D., Huwiler, A., et al. (2016). Upregulation of the S1P3 receptor in metastatic breast cancer cells increases migration and invasion by induction of PGE2 and EP2/EP4 activation. *Biochimica et Biophysica Acta*, *1861*, 1840–1851. https://doi.org/10.1016/j.bbalip.2016.09.005.

Fisher, K. E., Pop, A., Koh, W., Anthis, N. J., Saunders, W. B., & Davis, G. E. (2006). Tumor cell invasion of collagen matrices requires coordinate lipid agonist-induced G-protein and membrane-type matrix metalloproteinase-1-dependent signaling. *Molecular Cancer*, *5*, 69. https://doi.org/10.1186/1476-4598-5-69.

Fredriksson, R., Lagerstrom, M. C., Lundin, L. G., & Schioth, H. B. (2003). The G-protein-coupled receptors in the human genome form five main families. Phylogenetic analysis, paralogon groups, and fingerprints. *Molecular Pharmacology*, *63*, 1256–1272. https://doi.org/10.1124/mol.63.6.1256.

Fukuhara, S., Simmons, S., Kawamura, S., Inoue, A., Orba, Y., Tokudome, T., et al. (2012). The sphingosine-1-phosphate transporter Spns2 expressed on endothelial cells regulates lymphocyte trafficking in mice. *The Journal of Clinical Investigation*, *122*, 1416–1426. https://doi.org/10.1172/JCI60746.

Furusato, B., Mohamed, A., Uhlen, M., & Rhim, J. S. (2010). CXCR4 and cancer. *Pathology International*, *60*, 497–505. https://doi.org/10.1111/j.1440-1827.2010.02548.x.

Gao, D. Y., Lin Ts, T., Sung, Y. C., Liu, Y. C., Chiang, W. H., Chang, C. C., et al. (2015). CXCR4-targeted lipid-coated PLGA nanoparticles deliver sorafenib and overcome acquired drug resistance in liver cancer. *Biomaterials*, *67*, 194–203. https://doi.org/10.1016/j.biomaterials.2015.07.035.

Ghosh, E., Srivastava, A., Baidya, M., Kumari, P., Dwivedi, H., Nidhi, K., et al. (2017). A synthetic intrabody-based selective and generic inhibitor of GPCR endocytosis. *Nature Nanotechnology*, *12*, 1190–1198. https://doi.org/10.1038/nnano.2017.188.

Gil, M., Seshadri, M., Komorowski, M. P., Abrams, S. I., & Kozbor, D. (2013). Targeting CXCL12/CXCR4 signaling with oncolytic virotherapy disrupts tumor vasculature and inhibits breast cancer metastases. *Proceedings of the National Academy of Sciences of the United States of America*, *110*, E1291–E1300. https://doi.org/10.1073/pnas.1220580110.

Gill, D. M., & Meren, R. (1978). ADP-ribosylation of membrane proteins catalyzed by cholera toxin: Basis of the activation of adenylate cyclase. *Proceedings of the National Academy of Sciences of the United States of America*, *75*, 3050–3054. https://doi.org/10.1073/pnas.75.7.3050.

Giraudet, A.-L., Badel, J.-N., Cassier, P., et al. (2014). SYNFRIZZ-A phase Ia/Ib of a radiolabelled monoclonal AB for the treatment of relapsing synovial sarcoma. *Journal of Nuclear Medicine*, *55*, 223.

Gonda, K., Miyashita, M., Higuchi, H., Tada, H., Watanabe, T. M., Watanabe, M., et al. (2015). Predictive diagnosis of the risk of breast cancer recurrence after surgery by single-particle quantum dot imaging. *Scientific Reports*, *5*, 14322. https://doi.org/10.1038/srep14322.

Goodman, O. B., Jr., Krupnick, J. G., Santini, F., Gurevich, V. V., Penn, R. B., Gagnon, A. W., et al. (1996). Beta-arrestin acts as a clathrin adaptor in endocytosis of the beta2-adrenergic receptor. *Nature*, *383*, 447–450. https://doi.org/10.1038/383447a0.

Gratio, V., Beaufort, N., Seiz, L., Maier, J., Virca, G. D., Debela, M., et al. (2010). Kallikrein-related peptidase 4: A new activator of the aberrantly expressed protease-activated receptor 1 in colon cancer cells. *The American Journal of Pathology*, *176*, 1452–1461. https://doi.org/10.2353/ajpath.2010.090523.

Grisaru-Granovsky, S., Salah, Z., Maoz, M., Pruss, D., Beller, U., & Bar-Shavit, R. (2005). Differential expression of protease activated receptor 1 (Par1) and pY397FAK in benign and malignant human ovarian tissue samples. *International Journal of Cancer*, *113*, 372–378. https://doi.org/10.1002/ijc.20607.

Gurbel, P. A., Bliden, K. P., Turner, S. E., Tantry, U. S., Gesheff, M. G., Barr, T. P., et al. (2016). Cell-penetrating pepducin therapy targeting PAR1 in subjects with coronary artery disease. *Arteriosclerosis, Thrombosis, and Vascular Biology*, *36*, 189–197. https://doi.org/10.1161/ATVBAHA.115.306777.

Hait, N. C., Allegood, J., Maceyka, M., Strub, G. M., Harikumar, K. B., Singh, S. K., et al. (2009). Regulation of histone acetylation in the nucleus by sphingosine-1-phosphate. *Science*, *325*, 1254–1257. https://doi.org/10.1126/science.1176709.

Hauser, A. S., Attwood, M. M., Rask-Andersen, M., Schioth, H. B., & Gloriam, D. E. (2017). Trends in GPCR drug discovery: New agents, targets and indications. *Nature Reviews. Drug Discovery*, *16*, 829–842. https://doi.org/10.1038/nrd.2017.178.

Heider, I., Schulze, B., Oswald, E., Henklein, P., Scheele, J., & Kaufmann, R. (2004). PAR1-type thrombin receptor stimulates migration and matrix adhesion of human colon carcinoma cells by a PKCepsilon-dependent mechanism. *Oncology Research*, *14*, 475–482.

Hernandez, N. A., Correa, E., Avila, E. P., Vela, T. A., & Perez, V. M. (2009). PAR1 is selectively over expressed in high grade breast cancer patients: A cohort study. *Journal of Translational Medicine*, *7*, 47. https://doi.org/10.1186/1479-5876-7-47.

Herr, D. R., Reolo, M. J., Peh, Y. X., Wang, W., Lee, C. W., Rivera, R., et al. (2016). Sphingosine 1-phosphate receptor 2 (S1P2) attenuates reactive oxygen species formation and inhibits cell death: Implications for otoprotective therapy. *Scientific Reports*, *6*, 24541. https://doi.org/10.1038/srep24541.

Herr, D., Rodewald, M., Fraser, H. M., Hack, G., Konrad, R., Kreienberg, R., et al. (2008). Potential role of Renin-Angiotensin-system for tumor angiogenesis in receptor negative breast cancer. *Gynecologic Oncology*, *109*, 418–425. https://doi.org/10.1016/j.ygyno.2008.02.019.

Hilger, D., Masureel, M., & Kobilka, B. K. (2018). Structure and dynamics of GPCR signaling complexes. *Nature Structural & Molecular Biology, 25*, 4–12. https://doi.org/10.1038/s41594-017-0011-7.

Hirata, N., Yamada, S., Shoda, T., Kurihara, M., Sekino, Y., & Kanda, Y. (2014). Sphingosine-1-phosphate promotes expansion of cancer stem cells via S1PR3 by a ligand-independent Notch activation. *Nature Communications, 5*, 4806. https://doi.org/10.1038/ncomms5806.

Homan, K. T., & Tesmer, J. J. (2015). Molecular basis for small molecule inhibition of G protein-coupled receptor kinases. *ACS Chemical Biology, 10*, 246–256. https://doi.org/10.1021/cb5003976.

Hong, L., Li, Y., Liu, Q., Chen, Q., Chen, L., & Zhou, D. (1893). The Hippo signaling pathway in regenerative medicine. *Methods in Molecular Biology, 2019*, 353–370. https://doi.org/10.1007/978-1-4939-8910-2_26.

Hong, A. W., Meng, Z., Yuan, H. X., Plouffe, S. W., Moon, S., Kim, W., et al. (2017). Osmotic stress-induced phosphorylation by NLK at Ser128 activates YAP. *EMBO Reports, 18*, 72–86. https://doi.org/10.15252/embr.201642681.

Hu, W. M., Li, L., Jing, B. Q., Zhao, Y. S., Wang, C. L., Feng, L., et al. (2010). Effect of S1P5 on proliferation and migration of human esophageal cancer cells. *World Journal of Gastroenterology, 16*, 1859–1866. https://doi.org/10.3748/wjg.v16.i15.1859.

Hu, G. M., Mai, T. L., & Chen, C. M. (2017). Visualizing the GPCR network: Classification and evolution. *Scientific Reports, 7*, 15495. https://doi.org/10.1038/s41598-017-15707-9.

Hu, S., Wang, D., Wu, J., Jin, J., Wei, W., & Sun, W. (2013). Involvement of beta-arrestins in cancer progression. *Molecular Biology Reports, 40*, 1065–1071. https://doi.org/10.1007/s11033-012-2148-0.

Huang, Y. L., Chang, C. L., Tang, C. H., Lin, Y. C., Ju, T. K., Huang, W. P., et al. (2014). Extrinsic sphingosine 1-phosphate activates S1P5 and induces autophagy through generating endoplasmic reticulum stress in human prostate cancer PC-3 cells. *Cellular Signalling, 26*, 611–618. https://doi.org/10.1016/j.cellsig.2013.11.024.

Huang, M. M., Guo, A. B., Sun, J. F., Chen, X. L., & Yin, Z. Y. (2014). Angiotensin II promotes the progression of human gastric cancer. *Molecular Medicine Reports, 9*, 1056–1060. https://doi.org/10.3892/mmr.2014.1891.

Hutchings, C. J., Koglin, M., & Marshall, F. H. (2010). Therapeutic antibodies directed at G protein-coupled receptors. *mAbs, 2*, 594–606. https://doi.org/10.4161/mabs.2.6.13420.

Im, D. S., Heise, C. E., Harding, M. A., George, S. R., O'Dowd, B. F., Theodorescu, D., et al. (2000). Molecular cloning and characterization of a lysophosphatidic acid receptor, Edg-7, expressed in prostate. *Molecular Pharmacology, 57*, 753–759.

Iwanaga, T., Iwasaki, Y., Ohashi, M., Ohinata, R., Takahashi, K., Yamaguchi, T., et al. (2012). Inhibitory effect of CXCR4 blockers on a CXCR4-expressing gastric cancer cell line in nude mice. *Gan to Kagaku Ryoho, 39*, 1788–1790.

Jahan, I., Fujimoto, J., Alam, S. M., Sato, E., Sakaguchi, H., & Tamaya, T. (2007). Role of protease activated receptor-2 in tumor advancement of ovarian cancers. *Annals of Oncology, 18*, 1506–1512. https://doi.org/10.1093/annonc/mdm190.

Jahnichen, S., Blanchetot, C., Maussang, D., Gonzalez-Pajuelo, M., Chow, K. Y., Bosch, L., et al. (2010). CXCR4 nanobodies (VHH-based single variable domains) potently inhibit chemotaxis and HIV-1 replication and mobilize stem cells. *Proceedings of the National Academy of Sciences of the United States of America, 107*, 20565–20570. https://doi.org/10.1073/pnas.1012865107.

Jakubik, J., Bacakova, L., El-Fakahany, E. E., & Tucek, S. (1997). Positive cooperativity of acetylcholine and other agonists with allosteric ligands on muscarinic acetylcholine receptors. *Molecular Pharmacology, 52*, 172–179. https://doi.org/10.1124/mol.52.1.172.

Jeng, K. S., Jeng, C. J., Jeng, W. J., Sheen, I. S., Chang, C. F., Hsiau, H. I., et al. (2015). Sonic hedgehog pathway inhibitor mitigates mouse hepatocellular carcinoma. *American Journal of Surgery*, *210*, 554–560. https://doi.org/10.1016/j.amjsurg.2015.03.001.

Jenkinson, S., Thomson, M., McCoy, D., Edelstein, M., Danehower, S., Lawrence, W., et al. (2010). Blockade of X4-tropic HIV-1 cellular entry by GSK812397, a potent non-competitive CXCR4 receptor antagonist. *Antimicrobial Agents and Chemotherapy*, *54*, 817–824. https://doi.org/10.1128/AAC.01293-09.

Jeschke, G. (2012). DEER distance measurements on proteins. *Annual Review of Physical Chemistry*, *63*, 419–446. https://doi.org/10.1146/annurev-physchem-032511-143716.

Jin, E., Fujiwara, M., Pan, X., Ghazizadeh, M., Arai, S., Ohaki, Y., et al. (2003). Protease-activated receptor (PAR)-1 and PAR-2 participate in the cell growth of alveolar capillary endothelium in primary lung adenocarcinomas. *Cancer*, *97*, 703–713. https://doi.org/10.1002/cncr.11087.

Jo, E., Bhhatarai, B., Repetto, E., Guerrero, M., Riley, S., Brown, S. J., et al. (2012). Novel selective allosteric and bitopic ligands for the S1P(3) receptor. *ACS Chemical Biology*, *7*, 1975–1983. https://doi.org/10.1021/cb300392z.

Johnson, R., & Halder, G. (2014). The two faces of Hippo: targeting the Hippo pathway for regenerative medicine and cancer treatment. *Nature Reviews. Drug Discovery*, *13*, 63–79. https://doi.org/10.1038/nrd4161.

Joyal, J. S., Bhosle, V. K., & Chemtob, S. (2015). Subcellular G-protein coupled receptor signaling hints at greater therapeutic selectivity. *Expert Opinion on Therapeutic Targets*, *19*, 717–721. https://doi.org/10.1517/14728222.2015.1042365.

Kahsai, A. W., Wisler, J. W., Lee, J., Ahn, S., Cahill Iii, T. J., Dennison, S. M., et al. (2016). Conformationally selective RNA aptamers allosterically modulate the beta2-adrenoceptor. *Nature Chemical Biology*, *12*, 709–716. https://doi.org/10.1038/nchembio.2126.

Kang, D. S., Tian, X., & Benovic, J. L. (2014). Role of beta-arrestins and arrestin domain-containing proteins in G protein-coupled receptor trafficking. *Current Opinion in Cell Biology*, *27*, 63–71. https://doi.org/10.1016/j.ceb.2013.11.005.

Kaplon, H., & Reichert, J. M. (2019). Antibodies to watch in 2019. *mAbs*, *11*, 219–238. https://doi.org/10.1080/19420862.2018.1556465.

Katanaev, V. L., Ponzielli, R., Semeriva, M., & Tomlinson, A. (2005). Trimeric G protein-dependent frizzled signaling in Drosophila. *Cell*, *120*, 111–122. https://doi.org/10.1016/j.cell.2004.11.014.

Kato, Y., Kita, Y., Nishio, M., Hirasawa, Y., Ito, K., Yamanaka, T., et al. (1999). In vitro antiplatelet profile of FR171113, a novel non-peptide thrombin receptor antagonist. *European Journal of Pharmacology*, *384*, 197–202. https://doi.org/10.1016/s0014-2999(99)00658-5.

Kaufmann, R., Henklein, P., Henklein, P., & Settmacher, U. (2009). Green tea polyphenol epigallocatechin-3-gallate inhibits thrombin-induced hepatocellular carcinoma cell invasion and p42/p44-MAPKinase activation. *Oncology Reports*, *21*, 1261–1267. https://doi.org/10.3892/or_00000349.

Kaufmann, R., Rahn, S., Pollrich, K., Hertel, J., Dittmar, Y., Hommann, M., et al. (2007). Thrombin-mediated hepatocellular carcinoma cell migration: Cooperative action via proteinase-activated receptors 1 and 4. *Journal of Cellular Physiology*, *211*, 699–707. https://doi.org/10.1002/jcp.21027.

Kaur, G., Kim, J., Kaur, R., Tan, I., Bloch, O., Sun, M. Z., et al. (2013). G-protein coupled receptor kinase (GRK)-5 regulates proliferation of glioblastoma-derived stem cells. *Journal of Clinical Neuroscience*, *20*, 1014–1018. https://doi.org/10.1016/j.jocn.2012.10.008.

Kecskes, A., Tosh, D. K., Wei, Q., Gao, Z. G., & Jacobson, K. A. (2011). GPCR ligand dendrimer (GLiDe) conjugates: Adenosine receptor interactions of a series of multivalent xanthine antagonists. *Bioconjugate Chemistry, 22*, 1115–1127. https://doi.org/10.1021/bc1005812.

Kihara, Y., Maceyka, M., Spiegel, S., & Chun, J. (2014). Lysophospholipid receptor nomenclature review: IUPHAR review 8. *British Journal of Pharmacology, 171*, 3575–3594. https://doi.org/10.1111/bph.12678.

Kim, Y., Hechler, B., Gao, Z. G., Gachet, C., & Jacobson, K. A. (2009). PEGylated dendritic unimolecular micelles as versatile carriers for ligands of G protein-coupled receptors. *Bioconjugate Chemistry, 20*, 1888–1898. https://doi.org/10.1021/bc9001689.

Kim, E. S., Kim, J. S., Kim, S. G., Hwang, S., Lee, C. H., & Moon, A. (2011). Sphingosine 1-phosphate regulates matrix metalloproteinase-9 expression and breast cell invasion through S1P3-Galphaq coupling. *Journal of Cell Science, 124*, 2220–2230. https://doi.org/10.1242/jcs.076794.

Kim, Y., Klutz, A. M., Hechler, B., Gao, Z. G., Gachet, C., & Jacobson, K. A. (2009). Application of the functionalized congener approach to dendrimer-based signaling agents acting through A(2A) adenosine receptors. *Purinergic Signal, 5*, 39–50. https://doi.org/10.1007/s11302-008-9113-3.

Kim, G. S., Yang, L., Zhang, G., Zhao, H., Selim, M., McCullough, L. D., et al. (2015). Critical role of sphingosine-1-phosphate receptor-2 in the disruption of cerebrovascular integrity in experimental stroke. *Nature Communications, 6*, 7893. https://doi.org/10.1038/ncomms8893.

Kimura, T., Watanabe, T., Sato, K., Kon, J., Tomura, H., Tamama, K., et al. (2000). Sphingosine 1-phosphate stimulates proliferation and migration of human endothelial cells possibly through the lipid receptors, Edg-1 and Edg-3. *The Biochemical Journal, 348*(Pt. 1), 71–76.

Klarenbeek, A., Maussang, D., Blanchetot, C., Saunders, M., van der Woning, S., Smit, M., et al. (2012). Targeting chemokines and chemokine receptors with antibodies. *Drug Discovery Today: Technologies, 9*, e227–e314. https://doi.org/10.1016/j.ddtec.2012.05.003.

Klein Herenbrink, C., Sykes, D. A., Donthamsetti, P., Canals, M., Coudrat, T., Shonberg, J., et al. (2016). The role of kinetic context in apparent biased agonism at GPCRs. *Nature Communications,* 7, 10842. https://doi.org/10.1038/ncomms10842.

Kobilka, B. K., & Deupi, X. (2007). Conformational complexity of G-protein-coupled receptors. *Trends in Pharmacological Sciences, 28*, 397–406. https://doi.org/10.1016/j.tips.2007.06.003.

Koga, C., Kobayashi, S., Nagano, H., Tomimaru, Y., Hama, N., Wada, H., et al. (2014). Reprogramming using microRNA-302 improves drug sensitivity in hepatocellular carcinoma cells. *Annals of Surgical Oncology, 21*(Suppl. 4), S591–S600. https://doi.org/10.1245/s10434-014-3705-7.

Komachi, M., Sato, K., Tobo, M., Mogi, C., Yamada, T., Ohta, H., et al. (2012). Orally active lysophosphatidic acid receptor antagonist attenuates pancreatic cancer invasion and metastasis in vivo. *Cancer Science, 103*, 1099–1104. https://doi.org/10.1111/j.1349-7006.2012.02246.x.

Komolov, K. E., Du, Y., Duc, N. M., Betz, R. M., Rodrigues, J., Leib, R. D., et al. (2017). Structural and functional analysis of a beta2-adrenergic receptor complex with GRK5. *Cell, 169*, 407–421.e416. https://doi.org/10.1016/j.cell.2017.03.047.

Koontz, L. M., Liu-Chittenden, Y., Yin, F., Zheng, Y., Yu, J., Huang, B., et al. (2013). The Hippo effector Yorkie controls normal tissue growth by antagonizing scalloped-mediated default repression. *Developmental Cell, 25*, 388–401. https://doi.org/10.1016/j.devcel.2013.04.021.

Korinek, V., Barker, N., Morin, P. J., van Wichen, D., de Weger, R., Kinzler, K. W., et al. (1997). Constitutive transcriptional activation by a beta-catenin-Tcf complex in APC-/- colon carcinoma. *Science, 275*, 1784–1787. https://doi.org/10.1126/science.275.5307.1784.

Kotula, J. W., Sun, J., Li, M., Pratico, E. D., Fereshteh, M. P., Ahrens, D. P., et al. (2014). Targeted disruption of beta-arrestin 2-mediated signaling pathways by aptamer chimeras leads to inhibition of leukemic cell growth. *PLoS One, 9*, e93441. https://doi.org/10.1371/journal.pone.0093441.

Kovacs, J. J., Hara, M. R., Davenport, C. L., Kim, J., & Lefkowitz, R. J. (2009). Arrestin development: emerging roles for beta-arrestins in developmental signaling pathways. *Developmental Cell, 17*, 443–458. https://doi.org/10.1016/j.devcel.2009.09.011.

Kruse, A. C., Ring, A. M., Manglik, A., Hu, J., Hu, K., Eitel, K., et al. (2013). Activation and allosteric modulation of a muscarinic acetylcholine receptor. *Nature, 504*, 101–106. https://doi.org/10.1038/nature12735.

Kuhne, M. R., Mulvey, T., Belanger, B., Chen, S., Pan, C., Chong, C., et al. (2013). BMS-936564/MDX-1338: A fully human anti-CXCR4 antibody induces apoptosis in vitro and shows antitumor activity in vivo in hematologic malignancies. *Clinical Cancer Research, 19*, 357–366. https://doi.org/10.1158/1078-0432.CCR-12-2333.

Lappano, R., Jacquot, Y., & Maggiolini, M. (2018). GPCR modulation in breast cancer. *International Journal of Molecular Sciences, 19*, 3840. https://doi.org/10.3390/ijms19123840.

Le, P. N., McDermott, J. D., & Jimeno, A. (2015). Targeting the Wnt pathway in human cancers: Therapeutic targeting with a focus on OMP-54F28. *Pharmacology & Therapeutics, 146*, 1–11. https://doi.org/10.1016/j.pharmthera.2014.08.005.

Lebon, G., Warne, T., & Tate, C. G. (2012). Agonist-bound structures of G protein-coupled receptors. *Current Opinion in Structural Biology, 22*, 482–490. https://doi.org/10.1016/j.sbi.2012.03.007.

Lee, H. M., Lo, K. W., Wei, W., Tsao, S. W., Chung, G. T. Y., Ibrahim, M. H., et al. (2017). Oncogenic S1P signalling in EBV-associated nasopharyngeal carcinoma activates AKT and promotes cell migration through S1P receptor 3. *The Journal of Pathology, 242*, 62–72. https://doi.org/10.1002/path.4879.

Lee, M. J., Thangada, S., Claffey, K. P., Ancellin, N., Liu, C. H., Kluk, M., et al. (1999). Vascular endothelial cell adherens junction assembly and morphogenesis induced by sphingosine-1-phosphate. *Cell, 99*, 301–312. https://doi.org/10.1016/s0092-8674(00)81661-x.

Lei, Q. Y., Zhang, H., Zhao, B., Zha, Z. Y., Bai, F., Pei, X. H., et al. (2008). TAZ promotes cell proliferation and epithelial-mesenchymal transition and is inhibited by the hippo pathway. *Molecular and Cellular Biology, 28*, 2426–2436. https://doi.org/10.1128/MCB.01874-07.

Lepley, D., Paik, J. H., Hla, T., & Ferrer, F. (2005). The G protein-coupled receptor S1P2 regulates Rho/Rho kinase pathway to inhibit tumor cell migration. *Cancer Research, 65*, 3788–3795. https://doi.org/10.1158/0008-5472.CAN-04-2311.

Li, Y. P. (2013). GRK6 expression in patients with hepatocellular carcinoma. *Asian Pacific Journal of Tropical Medicine, 6*, 220–223. https://doi.org/10.1016/S1995-7645(13)60027-9.

Li, W., Ai, N., Wang, S., Bhattacharya, N., Vrbanac, V., Collins, M., et al. (2014). GRK3 is essential for metastatic cells and promotes prostate tumor progression. *Proceedings of the National Academy of Sciences of the United States of America, 111*, 1521–1526. https://doi.org/10.1073/pnas.1320638111.

Li, T. T., Alemayehu, M., Aziziyeh, A. I., Pape, C., Pampillo, M., Postovit, L. M., et al. (2009). Beta-arrestin/Ral signaling regulates lysophosphatidic acid-mediated migration and invasion of human breast tumor cells. *Molecular Cancer Research, 7*, 1064–1077. https://doi.org/10.1158/1541-7786.MCR-08-0578.

Li, Y., Liu, S., Ng, E. Y., Li, R., Poulsen, A., Hill, J., et al. (2018). Structural and ligand-binding analysis of the YAP-binding domain of transcription factor TEAD4. *The Biochemical Journal*, *475*, 2043–2055. https://doi.org/10.1042/BCJ20180225.

Li, M. H., Sanchez, T., Yamase, H., Hla, T., Oo, M. L., Pappalardo, A., et al. (2009). S1P/S1P1 signaling stimulates cell migration and invasion in Wilms tumor. *Cancer Letters*, *276*, 171–179. https://doi.org/10.1016/j.canlet.2008.11.025.

Li, X., Yao, W., Yuan, Y., Chen, P., Li, B., Li, J., et al. (2017). Targeting of tumour-infiltrating macrophages via CCL2/CCR2 signalling as a therapeutic strategy against hepatocellular carcinoma. *Gut*, *66*, 157–167. https://doi.org/10.1136/gutjnl-2015-310514.

Liang, Z., Zhan, W., Zhu, A., Yoon, Y., Lin, S., Sasaki, M., et al. (2012). Development of a unique small molecule modulator of CXCR4. *PLoS One*, *7*, e34038. https://doi.org/10.1371/journal.pone.0034038.

Lin, L., Han, M. M., Wang, F., Xu, L. L., Yu, H. X., & Yang, P. Y. (2014). CXCR7 stimulates MAPK signaling to regulate hepatocellular carcinoma progression. *Cell Death & Disease*, *5*, e1488. https://doi.org/10.1038/cddis.2014.392.

Lin, K. C., Moroishi, T., Meng, Z., Jeong, H. S., Plouffe, S. W., Sekido, Y., et al. (2017). Regulation of Hippo pathway transcription factor TEAD by p38 MAPK-induced cytoplasmic translocation. *Nature Cell Biology*, *19*, 996–1002. https://doi.org/10.1038/ncb3581.

Liu, X., Ahn, S., Kahsai, A. W., Meng, K. C., Latorraca, N. R., Pani, B., et al. (2017). Mechanism of intracellular allosteric beta2AR antagonist revealed by X-ray crystal structure. *Nature*, *548*, 480–484. https://doi.org/10.1038/nature23652.

Liu, Y., An, S., Ward, R., Yang, Y., Guo, X. X., Li, W., et al. (2016). G protein-coupled receptors as promising cancer targets. *Cancer Letters*, *376*, 226–239. https://doi.org/10.1016/j.canlet.2016.03.031.

Liu, J. Y., Chiang, T., Liu, C. H., Chern, G. G., Lin, T. T., Gao, D. Y., et al. (2015). Delivery of siRNA Using CXCR4-targeted nanoparticles modulates tumor microenvironment and achieves a potent antitumor response in liver cancer. *Molecular Therapy*, *23*, 1772–1782. https://doi.org/10.1038/mt.2015.147.

Liu, Y., Wada, R., Yamashita, T., Mi, Y., Deng, C. X., Hobson, J. P., et al. (2000). Edg-1, the G protein-coupled receptor for sphingosine-1-phosphate, is essential for vascular maturation. *The Journal of Clinical Investigation*, *106*, 951–961. https://doi.org/10.1172/JCI10905.

Lohse, M. J., Hein, P., Hoffmann, C., Nikolaev, V. O., Vilardaga, J. P., & Bunemann, M. (2008). Kinetics of G-protein-coupled receptor signals in intact cells. *British Journal of Pharmacology*, *153*(Suppl. 1), S125–S132. https://doi.org/10.1038/sj.bjp.0707656.

Long, J. S., Fujiwara, Y., Edwards, J., Tannahill, C. L., Tigyi, G., Pyne, S., et al. (2010). Sphingosine 1-phosphate receptor 4 uses HER2 (ERBB2) to regulate extracellular signal regulated kinase-1/2 in MDA-MB-453 breast cancer cells. *The Journal of Biological Chemistry*, *285*, 35957–35966. https://doi.org/10.1074/jbc.M110.117945.

Luo, J., & Yu, F. X. (2019). GPCR-hippo signaling in cancer. *Cell*, *8*, 426. https://doi.org/10.3390/cells8050426.

Ma, B., Chen, Y., Chen, L., Cheng, H., Mu, C., Li, J., et al. (2015). Hypoxia regulates Hippo signalling through the SIAH2 ubiquitin E3 ligase. *Nature Cell Biology*, *17*, 95–103. https://doi.org/10.1038/ncb3073.

Ma, B., Cheng, H., Gao, R., Mu, C., Chen, L., Wu, S., et al. (2016). Zyxin-Siah2-Lats2 axis mediates cooperation between Hippo and TGF-beta signalling pathways. *Nature Communications*, *7*, 11123. https://doi.org/10.1038/ncomms11123.

Ma, X., Xiong, Y., & Lee, L. T. O. (2018). Application of nanoparticles for targeting G protein-coupled receptors. *International Journal of Molecular Sciences*, *19*, 2006. https://doi.org/10.3390/ijms19072006.

Macfarlane, S. R., Seatter, M. J., Kanke, T., Hunter, G. D., & Plevin, R. (2001). Proteinase-activated receptors. *Pharmacological Reviews*, *53*, 245–282.
Maryanoff, B. E., Zhang, H. C., Andrade-Gordon, P., & Derian, C. K. (2003). Discovery of potent peptide-mimetic antagonists for the human thrombin receptor, protease-activated receptor-1 (PAR-1). *Current Medicinal Chemistry. Cardiovascular and Hematological Agents*, *1*, 13–36.
McNaughton, M., Pitman, M., Pitson, S. M., Pyne, N. J., & Pyne, S. (2016). Proteasomal degradation of sphingosine kinase 1 and inhibition of dihydroceramide desaturase by the sphingosine kinase inhibitors, SKi or ABC294640, induces growth arrest in androgen-independent LNCaP-AI prostate cancer cells. *Oncotarget*, 7, 16663–16675. https://doi.org/10.18632/oncotarget.7693.
Meng, Z., Qiu, Y., Lin, K. C., Kumar, A., Placone, J. K., Fang, C., et al. (2018). RAP2 mediates mechanoresponses of the Hippo pathway. *Nature*, *560*, 655–660. https://doi.org/10.1038/s41586-018-0444-0.
Metaye, T., Menet, E., Guilhot, J., & Kraimps, J. L. (2002). Expression and activity of G protein-coupled receptor kinases in differentiated thyroid carcinoma. *The Journal of Clinical Endocrinology and Metabolism*, *87*, 3279–3286. https://doi.org/10.1210/jcem.87.7.8618.
Miller, J., Agarwal, A., Devi, L. A., Fontanini, K., Hamilton, J. A., Pin, J. P., et al. (2009). Insider access: Pepducin symposium explores a new approach to GPCR modulation. *Annals of the New York Academy of Sciences*, *1180*(Suppl. 1), E1–12. https://doi.org/10.1111/j.1749-6632.2009.05326.x.
Miller, E., Yang, J., DeRan, M., Wu, C., Su, A. I., Bonamy, G. M., et al. (2012). Identification of serum-derived sphingosine-1-phosphate as a small molecule regulator of YAP. *Chemistry & Biology*, *19*, 955–962. https://doi.org/10.1016/j.chembiol.2012.07.005.
Mills, G. B., & Moolenaar, W. H. (2003). The emerging role of lysophosphatidic acid in cancer. *Nature Reviews. Cancer*, *3*, 582–591. https://doi.org/10.1038/nrc1143.
Mo, J. S., Yu, F. X., Gong, R., Brown, J. H., & Guan, K. L. (2012). Regulation of the Hippo-YAP pathway by protease-activated receptors (PARs). *Genes & Development*, *26*, 2138–2143. https://doi.org/10.1101/gad.197582.112.
Mujic-Delic, A., de Wit, R. H., Verkaar, F., & Smit, M. J. (2014). GPCR-targeting nanobodies: Attractive research tools, diagnostics, and therapeutics. *Trends in Pharmacological Sciences*, *35*, 247–255. https://doi.org/10.1016/j.tips.2014.03.003.
Muller, A., Homey, B., Soto, H., Ge, N., Catron, D., Buchanan, M. E., et al. (2001). Involvement of chemokine receptors in breast cancer metastasis. *Nature*, *410*, 50–56. https://doi.org/10.1038/35065016.
Murakami, T., Kumakura, S., Yamazaki, T., Tanaka, R., Hamatake, M., Okuma, K., et al. (2009). The novel CXCR4 antagonist KRH-3955 is an orally bioavailable and extremely potent inhibitor of human immunodeficiency virus type 1 infection: Comparative studies with AMD3100. *Antimicrobial Agents and Chemotherapy*, *53*, 2940–2948. https://doi.org/10.1128/AAC.01727-08.
Neves, S. R., Ram, P. T., & Iyengar, R. G. (2002). Protein pathways. *Science*, *296*, 1636–1639. https://doi.org/10.1126/science.1071550.
Nicholls, D. J., Tomkinson, N. P., Wiley, K. E., Brammall, A., Bowers, L., Grahames, C., et al. (2008). Identification of a putative intracellular allosteric antagonist binding-site in the CXC chemokine receptors 1 and 2. *Molecular Pharmacology*, 74, 1193–1202. https://doi.org/10.1124/mol.107.044610.
Nierodzik, M. L., Chen, K., Takeshita, K., Li, J. J., Huang, Y. Q., Feng, X. S., et al. (1998). Protease-activated receptor 1 (PAR-1) is required and rate-limiting for thrombin-enhanced experimental pulmonary metastasis. *Blood*, *92*, 3694–3700.

Nieto Gutierrez, A., & McDonald, P. H. (2018). GPCRs: Emerging anti-cancer drug targets. *Cellular Signalling*, *41*, 65–74. https://doi.org/10.1016/j.cellsig.2017.09.005.

Nishimura, A., Kitano, K., Takasaki, J., Taniguchi, M., Mizuno, N., Tago, K., et al. (2010). Structural basis for the specific inhibition of heterotrimeric Gq protein by a small molecule. *Proceedings of the National Academy of Sciences of the United States of America*, *107*, 13666–13671. https://doi.org/10.1073/pnas.1003553107.

Nishisho, I., Nakamura, Y., Miyoshi, Y., Miki, Y., Ando, H., Horii, A., et al. (1991). Mutations of chromosome 5q21 genes in FAP and colorectal cancer patients. *Science*, *253*, 665–669. https://doi.org/10.1126/science.1651563.

Nogues, L., Palacios-Garcia, J., Reglero, C., Rivas, V., Neves, M., Ribas, C., et al. (2018). G protein-coupled receptor kinases (GRKs) in tumorigenesis and cancer progression: GPCR regulators and signaling hubs. *Seminars in Cancer Biology*, *48*, 78–90. https://doi.org/10.1016/j.semcancer.2017.04.013.

Nogues, L., Reglero, C., Rivas, V., Neves, M., Penela, P., & Mayor, F., Jr. (2017). G-protein-coupled receptor kinase 2 as a potential modulator of the hallmarks of cancer. *Molecular Pharmacology*, *91*, 220–228. https://doi.org/10.1124/mol.116.107185.

Nogues, L., Reglero, C., Rivas, V., Salcedo, A., Lafarga, V., Neves, M., et al. (2016). G Protein-coupled receptor kinase 2 (GRK2) promotes breast tumorigenesis through a HDAC6-Pin1 axis. *eBioMedicine*, *13*, 132–145. https://doi.org/10.1016/j.ebiom.2016.09.030.

Noland, C. L., Gierke, S., Schnier, P. D., Murray, J., Sandoval, W. N., Sagolla, M., et al. (2016). Palmitoylation of TEAD transcription factors is required for their stability and function in Hippo pathway signaling. *Structure*, *24*, 179–186. https://doi.org/10.1016/j.str.2015.11.005.

Nugent, A., & Proia, R. L. (2017). The role of G protein-coupled receptors in lymphoid malignancies. *Cellular Signalling*, *39*, 95–107. https://doi.org/10.1016/j.cellsig.2017.08.002.

Nusse, R. (2005). Wnt signaling in disease and in development. *Cell Research*, *15*, 28–32. https://doi.org/10.1038/sj.cr.7290260.

Nusse, R., & Varmus, H. E. (1982). Many tumors induced by the mouse mammary tumor virus contain a provirus integrated in the same region of the host genome. *Cell*, *31*, 99–109. https://doi.org/10.1016/0092-8674(82)90409-3.

Nyunt, M. M., Becker, S., MacFarland, R. T., Chee, P., Scarborough, R., Everts, S., et al. (2008). Pharmacokinetic effect of AMD070, an Oral CXCR4 antagonist, on CYP3A4 and CYP2D6 substrates midazolam and dextromethorphan in healthy volunteers. *Journal of Acquired Immune Deficiency Syndromes*, *47*, 559–565. https://doi.org/10.1097/QAI.0b013e3181627566.

O'Boyle, G., Swidenbank, I., Marshall, H., Barker, C. E., Armstrong, J., White, S. A., et al. (2013). Inhibition of CXCR4-CXCL12 chemotaxis in melanoma by AMD11070. *British Journal of Cancer*, *108*, 1634–1640. https://doi.org/10.1038/bjc.2013.124.

O'Brien, N., Jones, S. T., Williams, D. G., Cunningham, H. B., Moreno, K., Visentin, B., et al. (2009). Production and characterization of monoclonal anti-sphingosine-1-phosphate antibodies. *Journal of Lipid Research*, *50*, 2245–2257. https://doi.org/10.1194/jlr.M900048-JLR200.

O'Callaghan, K., Kuliopulos, A., & Covic, L. (2012). Turning receptors on and off with intracellular pepducins: New insights into G-protein-coupled receptor drug development. *The Journal of Biological Chemistry*, *287*, 12787–12796. https://doi.org/10.1074/jbc.R112.355461.

Ochoa-Callejero, L., Perez-Martinez, L., Rubio-Mediavilla, S., Oteo, J. A., Martinez, A., & Blanco, J. R. (2013). Maraviroc, a CCR5 antagonist, prevents development of hepatocellular carcinoma in a mouse model. *PLoS One*, *8*, e53992. https://doi.org/10.1371/journal.pone.0053992.

Oh, E., Kim, J. Y., Cho, Y., An, H., Lee, N., Jo, H., et al. (2016). Overexpression of angiotensin II type 1 receptor in breast cancer cells induces epithelial-mesenchymal transition and promotes tumor growth and angiogenesis. *Biochimica et Biophysica Acta, 1863*, 1071–1081. https://doi.org/10.1016/j.bbamcr.2016.03.010.

O'Hayre, M., Salanga, C. L., Handel, T. M., & Allen, S. J. (2008). Chemokines and cancer: Migration, intracellular signalling and intercellular communication in the microenvironment. *Biochemical Journal, 409*, 635–649. https://doi.org/10.1042/BJ20071493.

Ohotski, J., Long, J. S., Orange, C., Elsberger, B., Mallon, E., Doughty, J., et al. (2012). Expression of sphingosine 1-phosphate receptor 4 and sphingosine kinase 1 is associated with outcome in oestrogen receptor-negative breast cancer. *British Journal of Cancer, 106*, 1453–1459. https://doi.org/10.1038/bjc.2012.98.

Onken, M. D., Makepeace, C. M., Kaltenbronn, K. M., Kanai, S. M., Todd, T. D., Wang, S., et al. (2018). Targeting nucleotide exchange to inhibit constitutively active G protein alpha subunits in cancer cells. *Science Signaling, 11*, eaao6852. https://doi.org/10.1126/scisignal.aao6852.

Oo, M. L., Thangada, S., Wu, M. T., Liu, C. H., Macdonald, T. L., Lynch, K. R., et al. (2007). Immunosuppressive and anti-angiogenic sphingosine 1-phosphate receptor-1 agonists induce ubiquitinylation and proteasomal degradation of the receptor. *The Journal of Biological Chemistry, 282*, 9082–9089. https://doi.org/10.1074/jbc.M610318200.

Ortiz Zacarias, N. V., Lenselink, E. B., IJzerman, A. P., Handel, T. M., & Heitman, L. H. (2018). Intracellular receptor modulation: Novel approach to target GPCRs. *Trends in Pharmacological Sciences, 39*, 547–559. https://doi.org/10.1016/j.tips.2018.03.002.

Oskeritzian, C. A., Hait, N. C., Wedman, P., Chumanevich, A., Kolawole, E. M., Price, M. M., et al. (2015). The sphingosine-1-phosphate/sphingosine-1-phosphate receptor 2 axis regulates early airway T-cell infiltration in murine mast cell-dependent acute allergic responses. *The Journal of Allergy and Clinical Immunology, 135*, 1008–1018.e1001. https://doi.org/10.1016/j.jaci.2014.10.044.

Oswald, C., Rappas, M., Kean, J., Dore, A. S., Errey, J. C., Bennett, K., et al. (2016). Intracellular allosteric antagonism of the CCR9 receptor. *Nature, 540*, 462–465. https://doi.org/10.1038/nature20606.

Otani, Y., Kijima, T., Kohmo, S., Oishi, S., Minami, T., Nagatomo, I., et al. (2012). Suppression of metastases of small cell lung cancer cells in mice by a peptidic CXCR4 inhibitor TF14016. *FEBS Letters, 586*, 3639–3644. https://doi.org/10.1016/j.febslet.2012.08.011.

Overholtzer, M., Zhang, J., Smolen, G. A., Muir, B., Li, W., Sgroi, D. C., et al. (2006). Transforming properties of YAP, a candidate oncogene on the chromosome 11q22 amplicon. *Proceedings of the National Academy of Sciences of the United States of America, 103*, 12405–12410. https://doi.org/10.1073/pnas.0605579103.

Pai, S. G., Carneiro, B. A., Mota, J. M., Costa, R., Leite, C. A., Barroso-Sousa, R., et al. (2017). Wnt/beta-catenin pathway: Modulating anticancer immune response. *Journal of Hematology & Oncology, 10*, 101. https://doi.org/10.1186/s13045-017-0471-6.

Paik, J. H., Chae, S., Lee, M. J., Thangada, S., & Hla, T. (2001). Sphingosine 1-phosphate-induced endothelial cell migration requires the expression of EDG-1 and EDG-3 receptors and Rho-dependent activation of alpha vbeta3- and beta1-containing integrins. *The Journal of Biological Chemistry, 276*, 11830–11837. https://doi.org/10.1074/jbc.M009422200.

Pan, D. (2010). The hippo signaling pathway in development and cancer. *Developmental Cell, 19*, 491–505. https://doi.org/10.1016/j.devcel.2010.09.011.

Parenti, M., Vigano, M. A., Newman, C. M., Milligan, G., & Magee, A. I. (1993). A novel N-terminal motif for palmitoylation of G-protein alpha subunits. *The Biochemical Journal, 291*(Pt. 2), 349–353. https://doi.org/10.1042/bj2910349.

Park, S. H., Das, B. B., Casagrande, F., Tian, Y., Nothnagel, H. J., Chu, M., et al. (2012). Structure of the chemokine receptor CXCR1 in phospholipid bilayers. *Nature*, *491*, 779–783. https://doi.org/10.1038/nature11580.

Park, H. W., & Guan, K. L. (2013). Regulation of the Hippo pathway and implications for anticancer drug development. *Trends in Pharmacological Sciences*, *34*, 581–589. https://doi.org/10.1016/j.tips.2013.08.006.

Patmanathan, S. N., Johnson, S. P., Lai, S. L., Panja Bernam, S., Lopes, V., Wei, W., et al. (2016). Aberrant expression of the S1P regulating enzymes, SPHK1 and SGPL1, contributes to a migratory phenotype in OSCC mediated through S1PR2. *Scientific Reports*, *6*, 25650. https://doi.org/10.1038/srep25650.

Paugh, S. W., Payne, S. G., Barbour, S. E., Milstien, S., & Spiegel, S. (2003). The immunosuppressant FTY720 is phosphorylated by sphingosine kinase type 2. *FEBS Letters*, *554*, 189–193. https://doi.org/10.1016/s0014-5793(03)01168-2.

Peled, A., Wald, O., & Burger, J. (2012). Development of novel CXCR4-based therapeutics. *Expert Opinion on Investigational Drugs*, *21*, 341–353. https://doi.org/10.1517/13543784.2012.656197.

Penela, P., Ribas, C., & Mayor, F., Jr. (2003). Mechanisms of regulation of the expression and function of G protein-coupled receptor kinases. *Cellular Signalling*, *15*, 973–981.

Peng, W. T., Sun, W. Y., Li, X. R., Sun, J. C., Du, J. J., & Wei, W. (2018). Emerging roles of G protein-coupled receptors in hepatocellular carcinoma. *International Journal of Molecular Sciences*, *19*, 1366. https://doi.org/10.3390/ijms19051366.

Perez, M., Lamothe, M., Maraval, C., Mirabel, E., Loubat, C., Planty, B., et al. (2009). Discovery of novel protease activated receptors 1 antagonists with potent antithrombotic activity in vivo. *Journal of Medicinal Chemistry*, *52*, 5826–5836. https://doi.org/10.1021/jm900553j.

Perez-Sayans, M., Somoza-Martin, J. M., Barros-Angueira, F., Gayoso-Diz, P., Otero-Rey, E. M., Gandra-Rey, J. M., et al. (2012). Activity of beta2-adrenergic receptor in oral squamous cell carcinoma is mediated by overexpression of the ADRBK2 gene: A pilot study. *Biotechnic & Histochemistry*, *87*, 179–186. https://doi.org/10.3109/10520295.2011.592151.

Pitcher, J. A., Freedman, N. J., & Lefkowitz, R. J. (1998). G protein-coupled receptor kinases. *Annual Review of Biochemistry*, *67*, 653–692. https://doi.org/10.1146/annurev.biochem.67.1.653.

Planesas, J. M., Perez-Nueno, V. I., Borrell, J. I., & Teixido, J. (2015). Studying the binding interactions of allosteric agonists and antagonists of the CXCR4 receptor. *Journal of Molecular Graphics & Modelling*, *60*, 1–14. https://doi.org/10.1016/j.jmgm.2015.05.004.

Ponnusamy, S., Selvam, S. P., Mehrotra, S., Kawamori, T., Snider, A. J., Obeid, L. M., et al. (2012). Communication between host organism and cancer cells is transduced by systemic sphingosine kinase 1/sphingosine 1-phosphate signalling to regulate tumour metastasis. *EMBO Molecular Medicine*, *4*, 761–775. https://doi.org/10.1002/emmm.201200244.

Pyne, N. J., & Pyne, S. (2010). Sphingosine 1-phosphate and cancer. *Nature Reviews. Cancer*, *10*, 489–503. https://doi.org/10.1038/nrc2875.

Pyne, N. J., & Pyne, S. (2011). Selectivity and specificity of sphingosine 1-phosphate receptor ligands: "Off-targets" or complex pharmacology? *Frontiers in Pharmacology*, *2*, 26. https://doi.org/10.3389/fphar.2011.00026.

Quoyer, J., Janz, J. M., Luo, J., Ren, Y., Armando, S., Lukashova, V., et al. (2013). Pepducin targeting the C-X-C chemokine receptor type 4 acts as a biased agonist favoring activation of the inhibitory G protein. *Proceedings of the National Academy of Sciences of the United States of America*, *110*, E5088–E5097. https://doi.org/10.1073/pnas.1312515110.

Ramachandran, R., & Hollenberg, M. D. (2008). Proteinases and signalling: pathophysiological and therapeutic implications via PARs and more. *British Journal of Pharmacology*, *153*(Suppl. 1), S263–S282. https://doi.org/10.1038/sj.bjp.0707507.

Ramsay, A. J., Reid, J. C., Adams, M. N., Samaratunga, H., Dong, Y., Clements, J. A., et al. (2008). Prostatic trypsin-like kallikrein-related peptidases (KLKs) and other prostate-expressed tryptic proteinases as regulators of signalling via proteinase-activated receptors (PARs). *Biological Chemistry*, *389*, 653–668. https://doi.org/10.1515/BC.2008.078.

Ranganathan, A., Dror, R. O., & Carlsson, J. (2014). Insights into the role of Asp79(2.50) in beta2 adrenergic receptor activation from molecular dynamics simulations. *Biochemistry*, *53*, 7283–7296. https://doi.org/10.1021/bi5008723.

Rankovic, Z., Brust, T. F., & Bohn, L. M. (2016). Biased agonism: An emerging paradigm in GPCR drug discovery. *Bioorganic & Medicinal Chemistry Letters*, *26*, 241–250. https://doi.org/10.1016/j.bmcl.2015.12.024.

Rask-Andersen, M., Masuram, S., & Schioth, H. B. (2014). The druggable genome: Evaluation of drug targets in clinical trials suggests major shifts in molecular class and indication. *Annual Review of Pharmacology and Toxicology*, *54*, 9–26. https://doi.org/10.1146/annurev-pharmtox-011613-135943.

Rasmussen, S. G., DeVree, B. T., Zou, Y., Kruse, A. C., Chung, K. Y., Kobilka, T. S., et al. (2011). Crystal structure of the beta2 adrenergic receptor-Gs protein complex. *Nature*, *477*, 549–555. https://doi.org/10.1038/nature10361.

Rasmussen, U. B., Vouret-Craviari, V., Jallat, S., Schlesinger, Y., Pages, G., Pavirani, A., et al. (1991). cDNA cloning and expression of a hamster alpha-thrombin receptor coupled to Ca2+ mobilization. *FEBS Letters*, *288*, 123–128. https://doi.org/10.1016/0014-5793(91)81017-3.

Ruiz-Gomez, A., & Mayor, F., Jr. (1997). Beta-adrenergic receptor kinase (GRK2) colocalizes with beta-adrenergic receptors during agonist-induced receptor internalization. *The Journal of Biological Chemistry*, *272*, 9601–9604. https://doi.org/10.1074/jbc.272.15.9601.

Salah, Z., Maoz, M., Pokroy, E., Lotem, M., Bar-Shavit, R., & Uziely, B. (2007). Protease-activated receptor-1 (hPar1), a survival factor eliciting tumor progression. *Molecular Cancer Research*, *5*, 229–240. https://doi.org/10.1158/1541-7786.MCR-06-0261.

Salomone, S., & Waeber, C. (2011). Selectivity and specificity of sphingosine-1-phosphate receptor ligands: Caveats and critical thinking in characterizing receptor-mediated effects. *Frontiers in Pharmacology*, *2*, 9. https://doi.org/10.3389/fphar.2011.00009.

Santos, R., Ursu, O., Gaulton, A., Bento, A. P., Donadi, R. S., Bologa, C. G., et al. (2017). A comprehensive map of molecular drug targets. *Nature Reviews. Drug Discovery*, *16*, 19–34. https://doi.org/10.1038/nrd.2016.230.

Satsu, H., Schaeffer, M. T., Guerrero, M., Saldana, A., Eberhart, C., Hodder, P., et al. (2013). A sphingosine 1-phosphate receptor 2 selective allosteric agonist. *Bioorganic & Medicinal Chemistry*, *21*, 5373–5382. https://doi.org/10.1016/j.bmc.2013.06.012.

Scarborough, R. M., Naughton, M. A., Teng, W., Hung, D. T., Rose, J., Vu, T. K., et al. (1992). Tethered ligand agonist peptides. Structural requirements for thrombin receptor activation reveal mechanism of proteolytic unmasking of agonist function. *The Journal of Biological Chemistry*, *267*, 13146–13149.

Serebruany, V. L., Kogushi, M., Dastros-Pitei, D., Flather, M., & Bhatt, D. L. (2009). The in-vitro effects of E5555, a protease-activated receptor (PAR)-1 antagonist, on platelet biomarkers in healthy volunteers and patients with coronary artery disease. *Thrombosis and Haemostasis*, *102*, 111–119. https://doi.org/10.1160/TH08-12-0805.

Shao, D., Zhai, P., Del Re, D. P., Sciarretta, S., Yabuta, N., Nojima, H., et al. (2014). A functional interaction between Hippo-YAP signalling and FoxO1 mediates the oxidative stress response. *Nature Communications*, *5*, 3315. https://doi.org/10.1038/ncomms4315.

Shearer, A. M., Rana, R., Austin, K., Baleja, J. D., Nguyen, N., Bohm, A., et al. (2016). Targeting liver fibrosis with a cell-penetrating protease-activated receptor-2 (PAR2) pepducin. *The Journal of Biological Chemistry*, *291*, 23188–23198. https://doi.org/10.1074/jbc.M116.732743.

Shukla, A. K. (2014). Biasing GPCR signaling from inside. *Science Signaling*, 7, pe3. https://doi.org/10.1126/scisignal.2005021.

Shukla, A. K., Westfield, G. H., Xiao, K., Reis, R. I., Huang, L. Y., Tripathi-Shukla, P., et al. (2014). Visualization of arrestin recruitment by a G-protein-coupled receptor. *Nature*, *512*, 218–222. https://doi.org/10.1038/nature13430.

Smith, J. S., Lefkowitz, R. J., & Rajagopal, S. (2018). Biased signalling: From simple switches to allosteric microprocessors. *Nature Reviews. Drug Discovery*, *17*, 243–260. https://doi.org/10.1038/nrd.2017.229.

Smith, D. C., Rosen, L. S., Chugh, R., et al. (2013). First-in human evaluation of the human monoclonal antibody vantictumab (OMP-18R5; anti-Frizzled) targeting the WNT pathway in a phase I study for patients with advanced solid tumors. *Journal of Clinical Oncology*, *31*, 2540.

Sorkin, A., & von Zastrow, M. (2009). Endocytosis and signalling: Intertwining molecular networks. *Nature Reviews. Molecular Cell Biology*, *10*, 609–622. https://doi.org/10.1038/nrm2748.

Sorrentino, G., Ruggeri, N., Specchia, V., Cordenonsi, M., Mano, M., Dupont, S., et al. (2014). Metabolic control of YAP and TAZ by the mevalonate pathway. *Nature Cell Biology*, *16*, 357–366. https://doi.org/10.1038/ncb2936.

Spiegel, S., & Milstien, S. (2007). Functions of the multifaceted family of sphingosine kinases and some close relatives. *The Journal of Biological Chemistry*, *282*, 2125–2129. https://doi.org/10.1074/jbc.R600028200.

Srivastava, A., Gupta, B., Gupta, C., & Shukla, A. K. (2015). Emerging functional divergence of beta-arrestin isoforms in GPCR function. *Trends in Endocrinology and Metabolism*, *26*, 628–642. https://doi.org/10.1016/j.tem.2015.09.001.

Stanger, B. Z. (2012). Quit your YAPing: A new target for cancer therapy. *Genes & Development*, *26*, 1263–1267. https://doi.org/10.1101/gad.196501.112.

Staus, D. P., Wingler, L. M., Strachan, R. T., Rasmussen, S. G., Pardon, E., Ahn, S., et al. (2014). Regulation of beta2-adrenergic receptor function by conformationally selective single-domain intrabodies. *Molecular Pharmacology*, *85*, 472–481. https://doi.org/10.1124/mol.113.089516.

Stone, N. D., Dunaway, S. B., Flexner, C., Tierney, C., Calandra, G. B., Becker, S., et al. (2007). Multiple-dose escalation study of the safety, pharmacokinetics, and biologic activity of oral AMD070, a selective CXCR4 receptor inhibitor, in human subjects. *Antimicrobial Agents and Chemotherapy*, *51*, 2351–2358. https://doi.org/10.1128/AAC.00013-07.

Strange, P. G. (2008). Signaling mechanisms of GPCR ligands. *Current Opinion in Drug Discovery & Development*, *11*, 196–202.

Striedinger, K., VandenBerg, S. R., Baia, G. S., McDermott, M. W., Gutmann, D. H., & Lal, A. (2008). The neurofibromatosis 2 tumor suppressor gene product, merlin, regulates human meningioma cell growth by signaling through YAP. *Neoplasia*, *10*, 1204–1212. https://doi.org/10.1593/neo.08642.

Strub, G. M., Paillard, M., Liang, J., Gomez, L., Allegood, J. C., Hait, N. C., et al. (2011). Sphingosine-1-phosphate produced by sphingosine kinase 2 in mitochondria interacts with prohibitin 2 to regulate complex IV assembly and respiration. *The FASEB Journal*, *25*, 600–612. https://doi.org/10.1096/fj.10-167502.

Su, S., Li, Y., Luo, Y., Sheng, Y., Su, Y., Padia, R. N., et al. (2009). Proteinase-activated receptor 2 expression in breast cancer and its role in breast cancer cell migration. *Oncogene*, *28*, 3047–3057. https://doi.org/10.1038/onc.2009.163.

Sugimoto, N., Takuwa, N., Okamoto, H., Sakurada, S., & Takuwa, Y. (2003). Inhibitory and stimulatory regulation of Rac and cell motility by the G12/13-Rho and Gi pathways integrated downstream of a single G protein-coupled sphingosine-1-phosphate receptor isoform. *Molecular and Cellular Biology*, *23*, 1534–1545. https://doi.org/10.1128/mcb.23.5.1534-1545.2003.

Tabata, K., Baba, K., Shiraishi, A., Ito, M., & Fujita, N. (2007). The orphan GPCR GPR87 was deorphanized and shown to be a lysophosphatidic acid receptor. *Biochemical and Biophysical Research Communications*, *363*, 861–866. https://doi.org/10.1016/j.bbrc.2007.09.063.

Takabe, K., Paugh, S. W., Milstien, S., & Spiegel, S. (2008). "Inside-out" signaling of sphingosine-1-phosphate: Therapeutic targets. *Pharmacological Reviews*, *60*, 181–195. https://doi.org/10.1124/pr.107.07113.

Tamamura, H., Hiramatsu, K., Ueda, S., Wang, Z., Kusano, S., Terakubo, S., et al. (2005). Stereoselective synthesis of [L-Arg-L/D-3-(2-naphthyl)alanine]-type (E)-alkene dipeptide isosteres and its application to the synthesis and biological evaluation of pseudopeptide analogues of the CXCR4 antagonist FC131. *Journal of Medicinal Chemistry*, *48*, 380–391. https://doi.org/10.1021/jm049429h.

Tamamura, H., Hori, A., Kanzaki, N., Hiramatsu, K., Mizumoto, M., Nakashima, H., et al. (2003). T140 analogs as CXCR4 antagonists identified as anti-metastatic agents in the treatment of breast cancer. *FEBS Letters*, *550*, 79–83. https://doi.org/10.1016/s0014-5793(03)00824-x.

Tello-Montoliu, A., Tomasello, S. D., Ueno, M., & Angiolillo, D. J. (2011). Antiplatelet therapy: Thrombin receptor antagonists. *British Journal of Clinical Pharmacology*, *72*, 658–671. https://doi.org/10.1111/j.1365-2125.2010.03884.x.

Tesmer, V. M., Lennarz, S., Mayer, G., & Tesmer, J. J. (2012). Molecular mechanism for inhibition of g protein-coupled receptor kinase 2 by a selective RNA aptamer. *Structure*, *20*, 1300–1309. https://doi.org/10.1016/j.str.2012.05.002.

Thal, D. M., Sun, B., Feng, D., Nawaratne, V., Leach, K., Felder, C. C., et al. (2016). Crystal structures of the M1 and M4 muscarinic acetylcholine receptors. *Nature*, *531*, 335–340. https://doi.org/10.1038/nature17188.

Thal, D. M., Yeow, R. Y., Schoenau, C., Huber, J., & Tesmer, J. J. (2011). Molecular mechanism of selectivity among G protein-coupled receptor kinase 2 inhibitors. *Molecular Pharmacology*, *80*, 294–303. https://doi.org/10.1124/mol.111.071522.

Tobin, A. B. (2008). G-protein-coupled receptor phosphorylation: Where, when and by whom. *British Journal of Pharmacology*, *153*(Suppl. 1), S167–S176. https://doi.org/10.1038/sj.bjp.0707662.

Tressel, S. L., Koukos, G., Tchernychev, B., Jacques, S. L., Covic, L., & Kuliopulos, A. (2011). Pharmacology, biodistribution, and efficacy of GPCR-based pepducins in disease models. *Methods in Molecular Biology*, *683*, 259–275. https://doi.org/10.1007/978-1-60761-919-2_19.

Urbano, M., Guerrero, M., Velaparthi, S., Crisp, M., Chase, P., Hodder, P., et al. (2011). Discovery, synthesis and SAR analysis of novel selective small molecule S1P4-R agonists based on a (2Z,5Z)-5-((pyrrol-3-yl)methylene)-3-alkyl-2-(alkylimino)thiazolidin-4-one chemotype. *Bioorganic & Medicinal Chemistry Letters*, *21*, 6739–6745. https://doi.org/10.1016/j.bmcl.2011.09.049.

Valenta, T., Hausmann, G., & Basler, K. (2012). The many faces and functions of beta-catenin. *The EMBO Journal*, *31*, 2714–2736. https://doi.org/10.1038/emboj.2012.150.

van Amerongen, R. (2012). Alternative Wnt pathways and receptors. *Cold Spring Harbor Perspectives in Biology*, *4*, a007914. https://doi.org/10.1101/cshperspect.a007914.

van der Westhuizen, E. T., Valant, C., Sexton, P. M., & Christopoulos, A. (2015). Endogenous allosteric modulators of G protein-coupled receptors. *The Journal of Pharmacology and Experimental Therapeutics*, *353*, 246–260. https://doi.org/10.1124/jpet.114.221606.

Venant, H., Rahmaniyan, M., Jones, E. E., Lu, P., Lilly, M. B., Garrett-Mayer, E., et al. (2015). The sphingosine kinase 2 inhibitor ABC294640 reduces the growth of prostate cancer cells and results in accumulation of dihydroceramides in vitro and in vivo. *Molecular Cancer Therapeutics*, *14*, 2744–2752. https://doi.org/10.1158/1535-7163.MCT-15-0279.

Versteeg, H. H., Schaffner, F., Kerver, M., Ellies, L. G., Andrade-Gordon, P., Mueller, B. M., et al. (2008). Protease-activated receptor (PAR) 2, but not PAR1, signaling promotes the development of mammary adenocarcinoma in polyoma middle T mice. *Cancer Research*, *68*, 7219–7227. https://doi.org/10.1158/0008-5472.CAN-08-0419.

Violin, J. D., Crombie, A. L., Soergel, D. G., & Lark, M. W. (2014). Biased ligands at G-protein-coupled receptors: Promise and progress. *Trends in Pharmacological Sciences*, *35*, 308–316. https://doi.org/10.1016/j.tips.2014.04.007.

Visentin, B., Vekich, J. A., Sibbald, B. J., Cavalli, A. L., Moreno, K. M., Matteo, R. G., et al. (2006). Validation of an anti-sphingosine-1-phosphate antibody as a potential therapeutic in reducing growth, invasion, and angiogenesis in multiple tumor lineages. *Cancer Cell*, *9*, 225–238. https://doi.org/10.1016/j.ccr.2006.02.023.

Vu, T. K., Hung, D. T., Wheaton, V. I., & Coughlin, S. R. (1991). Molecular cloning of a functional thrombin receptor reveals a novel proteolytic mechanism of receptor activation. *Cell*, *64*, 1057–1068. https://doi.org/10.1016/0092-8674(91)90261-v.

Waldschmidt, H. V., Homan, K. T., Cruz-Rodriguez, O., Cato, M. C., Waninger-Saroni, J., Larimore, K. M., et al. (2016). Structure-based design, synthesis, and biological evaluation of highly selective and potent G protein-coupled receptor kinase 2 inhibitors. *Journal of Medicinal Chemistry*, *59*, 3793–3807. https://doi.org/10.1021/acs.jmedchem.5b02000.

Wang, Z., Liu, P., Zhou, X., Wang, T., Feng, X., Sun, Y. P., et al. (2017). Endothelin promotes colorectal tumorigenesis by activating YAP/TAZ. *Cancer Research*, *77*, 2413–2423. https://doi.org/10.1158/0008-5472.CAN-16-3229.

Watari, K., Nakaya, M., & Kurose, H. (2014). Multiple functions of G protein-coupled receptor kinases. *Journal of Molecular Signaling*, *9*, 1. https://doi.org/10.1186/1750-2187-9-1.

Watson, C., Long, J. S., Orange, C., Tannahill, C. L., Mallon, E., McGlynn, L. M., et al. (2010). High expression of sphingosine 1-phosphate receptors, S1P1 and S1P3, sphingosine kinase 1, and extracellular signal-regulated kinase-1/2 is associated with development of tamoxifen resistance in estrogen receptor-positive breast cancer patients. *The American Journal of Pathology*, *177*, 2205–2215. https://doi.org/10.2353/ajpath.2010.100220.

Webb, D. R., Handel, T. M., Kretz-Rommel, A., & Stevens, R. C. (2013). Opportunities for functional selectivity in GPCR antibodies. *Biochemical Pharmacology*, *85*, 147–152. https://doi.org/10.1016/j.bcp.2012.08.021.

Weekes, C., Berlin, J., Lenz, H.-J., et al. (2016). Phase 1b study of WNT inhibitor ipafricept (IPA, decoy receptor for WNT ligands) with nab-paclitaxel (Nab-P) and gemcitabine (G) in patients (pts) with previously untreated stage IV pancreatic cancer (PC). *Annals of Oncology*, *27*, 367PD. https://doi.org/10.1093/annonc/mdw368.10.

Wei, Z., Hurtt, R., Gu, T., Bodzin, A. S., Koch, W. J., & Doria, C. (2013). GRK2 negatively regulates IGF-1R signaling pathway and cyclins' expression in HepG2 cells. *Journal of Cellular Physiology*, *228*, 1897–1901. https://doi.org/10.1002/jcp.24353.

Weiss, R. A. (2001). Gulliver's travels in HIVland. *Nature*, *410*, 963–967. https://doi.org/10.1038/35073632.

Wendt, E., & Keshav, S. (2015). CCR9 antagonism: Potential in the treatment of inflammatory bowel disease. *Clinical and Experimental Gastroenterology*, *8*, 119–130. https://doi.org/10.2147/CEG.S48305.

Wennmann, D. O., Vollenbroker, B., Eckart, A. K., Bonse, J., Erdmann, F., Wolters, D. A., et al. (2014). The Hippo pathway is controlled by angiotensin II signaling and its reactivation induces apoptosis in podocytes. *Cell Death & Disease*, *5*, e1519. https://doi.org/10.1038/cddis.2014.476.

West, R. E., Jr., Moss, J., Vaughan, M., Liu, T., & Liu, T. Y. (1985). Pertussis toxin-catalyzed ADP-ribosylation of transducin. Cysteine 347 is the ADP-ribose acceptor site. *The Journal of Biological Chemistry*, *260*, 14428–14430.

White, C., Alshaker, H., Cooper, C., Winkler, M., & Pchejetski, D. (2016). The emerging role of FTY720 (Fingolimod) in cancer treatment. *Oncotarget*, 7, 23106–23127. https://doi.org/10.18632/oncotarget.7145.

Wishart, D. S. (2011). Interpreting protein chemical shift data. *Progress in Nuclear Magnetic Resonance Spectroscopy*, *58*, 62–87. https://doi.org/10.1016/j.pnmrs.2010.07.004.

Wootten, D., Christopoulos, A., & Sexton, P. M. (2013). Emerging paradigms in GPCR allostery: Implications for drug discovery. *Nature Reviews. Drug Discovery*, *12*, 630–644. https://doi.org/10.1038/nrd4052.

Wu, H., Wei, L., Fan, F., Ji, S., Zhang, S., Geng, J., et al. (2015). Integration of Hippo signalling and the unfolded protein response to restrain liver overgrowth and tumorigenesis. *Nature Communications*, *6*, 6239. https://doi.org/10.1038/ncomms7239.

Xiao, K., & Sun, J. (2018). Elucidating structural and molecular mechanisms of beta-arrestin-biased agonism at GPCRs via MS-based proteomics. *Cellular Signalling*, *41*, 56–64. https://doi.org/10.1016/j.cellsig.2017.09.013.

Xue, L., Rovira, X., Scholler, P., Zhao, H., Liu, J., Pin, J. P., et al. (2015). Major ligand-induced rearrangement of the heptahelical domain interface in a GPCR dimer. *Nature Chemical Biology*, *11*, 134–140. https://doi.org/10.1038/nchembio.1711.

Yamaguchi, H., Kitayama, J., Takuwa, N., Arikawa, K., Inoki, I., Takehara, K., et al. (2003). Sphingosine-1-phosphate receptor subtype-specific positive and negative regulation of Rac and haematogenous metastasis of melanoma cells. *The Biochemical Journal*, *374*, 715–722. https://doi.org/10.1042/BJ20030381.

Yamamoto, S., Yako, Y., Fujioka, Y., Kajita, M., Kameyama, T., Kon, S., et al. (2016). A role of the sphingosine-1-phosphate (S1P)-S1P receptor 2 pathway in epithelial defense against cancer (EDAC). *Molecular Biology of the Cell*, *27*, 491–499. https://doi.org/10.1091/mbc.E15-03-0161.

Yamamura, S., Hakomori, S., Wada, A., & Igarashi, Y. (2000). Sphingosine-1-phosphate inhibits haptotactic motility by overproduction of focal adhesion sites in B16 melanoma cells through EDG-induced activation of Rho. *Annals of the New York Academy of Sciences*, *905*, 301–307. https://doi.org/10.1111/j.1749-6632.2000.tb06566.x.

Yamashita, H., Kitayama, J., Shida, D., Yamaguchi, H., Mori, K., Osada, M., et al. (2006). Sphingosine 1-phosphate receptor expression profile in human gastric cancer cells: Differential regulation on the migration and proliferation. *The Journal of Surgical Research*, *130*, 80–87. https://doi.org/10.1016/j.jss.2005.08.004.

Yang, E., Boire, A., Agarwal, A., Nguyen, N., O'Callaghan, K., Tu, P., et al. (2009). Blockade of PAR1 signaling with cell-penetrating pepducins inhibits Akt survival pathways in breast cancer cells and suppresses tumor survival and metastasis. *Cancer Research*, *69*, 6223–6231. https://doi.org/10.1158/0008-5472.CAN-09-0187.

Yang, E., Cisowski, J., Nguyen, N., O'Callaghan, K., Xu, J., Agarwal, A., et al. (2016). Dysregulated protease activated receptor 1 (PAR1) promotes metastatic phenotype in breast cancer through HMGA2. *Oncogene*, *35*, 1529–1540. https://doi.org/10.1038/onc.2015.217.

Yang, M., Zhong, W. W., Srivastava, N., Slavin, A., Yang, J., Hoey, T., et al. (2005). G protein-coupled lysophosphatidic acid receptors stimulate proliferation of colon cancer cells through the {beta}-catenin pathway. *Proceedings of the National Academy of Sciences of the United States of America*, *102*, 6027–6032. https://doi.org/10.1073/pnas.0501535102.

Yoshikawa, Y., Kobayashi, K., Oishi, S., Fujii, N., & Furuya, T. (2012). Molecular modeling study of cyclic pentapeptide CXCR4 antagonists: New insight into CXCR4-FC131 interactions. *Bioorganic & Medicinal Chemistry Letters*, *22*, 2146–2150. https://doi.org/10.1016/j.bmcl.2012.01.134.

Young, N., Pearl, D. K., & Van Brocklyn, J. R. (2009). Sphingosine-1-phosphate regulates glioblastoma cell invasiveness through the urokinase plasminogen activator system and CCN1/Cyr61. *Molecular Cancer Research*, *7*, 23–32. https://doi.org/10.1158/1541-7786.MCR-08-0061.

Young, N., & Van Brocklyn, J. R. (2007). Roles of sphingosine-1-phosphate (S1P) receptors in malignant behavior of glioma cells. Differential effects of S1P2 on cell migration and invasiveness. *Experimental Cell Research*, *313*, 1615–1627. https://doi.org/10.1016/j.yexcr.2007.02.009.

Young, D., Waitches, G., Birchmeier, C., Fasano, O., & Wigler, M. (1986). Isolation and characterization of a new cellular oncogene encoding a protein with multiple potential transmembrane domains. *Cell*, *45*, 711–719. https://doi.org/10.1016/0092-8674(86)90785-3.

Yu, O. M., Miyamoto, S., & Brown, J. H. (2016). Myocardin-related transcription factor A and Yes-associated protein exert dual control in G protein-coupled receptor- and RhoA-mediated transcriptional regulation and cell proliferation. *Molecular and Cellular Biology*, *36*, 39–49. https://doi.org/10.1128/MCB.00772-15.

Yu, F. X., Mo, J. S., & Guan, K. L. (2012). Upstream regulators of the Hippo pathway. *Cell Cycle*, *11*, 4097–4098. https://doi.org/10.4161/cc.22322.

Yu, F. X., Zhao, B., Panupinthu, N., Jewell, J. L., Lian, I., Wang, L. H., et al. (2012). Regulation of the Hippo-YAP pathway by G-protein-coupled receptor signaling. *Cell*, *150*, 780–791. https://doi.org/10.1016/j.cell.2012.06.037.

Yun, C. C., Sun, H., Wang, D., Rusovici, R., Castleberry, A., Hall, R. A., et al. (2005). LPA2 receptor mediates mitogenic signals in human colon cancer cells. *American Journal of Physiology. Cell Physiology*, *289*, C2–11. https://doi.org/10.1152/ajpcell.00610.2004.

Zhang, H. C., Derian, C. K., Andrade-Gordon, P., Hoekstra, W. J., McComsey, D. F., White, K. B., et al. (2001). Discovery and optimization of a novel series of thrombin receptor (par-1) antagonists: Potent, selective peptide mimetics based on indole and indazole templates. *Journal of Medicinal Chemistry*, *44*, 1021–1024. https://doi.org/10.1021/jm000506s.

Zhang, W., Gao, Y., Li, P., Shi, Z., Guo, T., Li, F., et al. (2014). VGLL4 functions as a new tumor suppressor in lung cancer by negatively regulating the YAP-TEAD transcriptional complex. *Cell Research*, *24*, 331–343. https://doi.org/10.1038/cr.2014.10.

Zhao, Q., Ishibashi, M., Hiasa, K., Tan, C., Takeshita, A., & Egashira, K. (2004). Essential role of vascular endothelial growth factor in angiotensin II-induced vascular inflammation and remodeling. *Hypertension*, *44*, 264–270. https://doi.org/10.1161/01.HYP.0000138688.78906.6b.

Zhao, B., Li, L., Tumaneng, K., Wang, C. Y., & Guan, K. L. (2010). A coordinated phosphorylation by Lats and CK1 regulates YAP stability through SCF(beta-TRCP). *Genes & Development*, *24*, 72–85. https://doi.org/10.1101/gad.1843810.

Zhao, B., Wei, X., Li, W., Udan, R. S., Yang, Q., Kim, J., et al. (2007). Inactivation of YAP oncoprotein by the Hippo pathway is involved in cell contact inhibition and tissue growth control. *Genes & Development*, *21*, 2747–2761. https://doi.org/10.1101/gad.1602907.

Zheng, Y., Qin, L., Zacarias, N. V., de Vries, H., Han, G. W., Gustavsson, M., et al. (2016). Structure of CC chemokine receptor 2 with orthosteric and allosteric antagonists. *Nature*, *540*, 458–461. https://doi.org/10.1038/nature20605.

Zhou, L., Luo, Y., Sato, S., Tanabe, E., Kitayoshi, M., Fujiwara, R., et al. (2014). Role of two types of angiotensin II receptors in colorectal carcinoma progression. *Pathobiology*, *81*, 169–175. https://doi.org/10.1159/000362092.

Zhu, H., Cheng, X., Niu, X., Zhang, Y., Guan, J., Liu, X., et al. (2015). Proton-sensing GPCR-YAP signalling promotes cell proliferation and survival. *International Journal of Biological Sciences*, *11*, 1181–1189. https://doi.org/10.7150/ijbs.12500.

Zlotnik, A., Burkhardt, A. M., & Homey, B. (2011). Homeostatic chemokine receptors and organ-specific metastasis. *Nature Reviews. Immunology*, *11*, 597–606. https://doi.org/10.1038/nri3049.

CHAPTER FOUR

The diverse and complex roles of atypical chemokine receptors in cancer: From molecular biology to clinical relevance and therapy

Elin Sjöberg[a,†], Max Meyrath[b,†], Andy Chevigné[b], Arne Östman[a], Martin Augsten[c,*,‡], Martyna Szpakowska[b,*,‡]
[a]Department of Oncology-Pathology, Karolinska Institutet, Stockholm, Sweden
[b]Department of Infection and Immunity, Immuno-Pharmacology and Interactomics, Luxembourg Institute of Health (LIH), Esch-sur-Alzette, Luxembourg
[c]amcure GmbH, Eggenstein-Leopoldshafen, Germany
*Corresponding authors: e-mail address: augsten@gmx.net; martyna.szpakowska@lih.lu

Contents

† Co-first authors contributed equally.
‡ Co-last authors contributed equally.

Advances in Cancer Research, Volume 145
ISSN 0065-230X
https://doi.org/10.1016/bs.acr.2019.12.001

Abstract

Chemokines regulate directed cell migration, proliferation and survival and are key components in cancer biology. They exert their functions by interacting with seven-transmembrane domain receptors that signal through G proteins (GPCRs). A subgroup of four chemokine receptors known as the atypical chemokine receptors (ACKRs) has emerged as essential regulators of the chemokine functions. ACKRs play diverse and complex roles in tumor biology from tumor initiation to metastasis, including cancer cell proliferation, adherence to endothelium, epithelial-mesenchymal transition (EMT), extravasation from blood vessels, tumor-associated angiogenesis or protection from immunological responses. This chapter gives an overview on the established and emerging roles that the atypical chemokine receptors ACKR1, ACKR2, ACKR3 and ACKR4 play in the different phases of cancer development and dissemination, their clinical relevance, as well as on the hurdles to overcome in ACKRs targeting as cancer therapy.

1. Introduction

Chemokines, or chemotactic cytokines, regulate directed cell migration, proliferation and survival and are key components in physiology and pathological processes including cancer (Zlotnik & Yoshie, 2012). They act through seven-transmembrane domain G protein-coupled receptors (GPCRs). Functionally, chemokines can be divided into molecules with homeostatic or inflammatory properties. Structurally, on the basis of a specific cysteine motif, chemokines are classified into CC, CXC, XC and CX3C subfamilies and accordingly bind to their respective chemokine receptor subfamilies CCR, CXCR, XCR and CX3CR (Zlotnik & Yoshie, 2000). Over the last years, a new family of four chemokine receptors has emerged as important regulators of chemokine functions. Formerly called chemokine-binding proteins, decoys, scavengers or interceptors, the standard nomenclature for this family of receptors is now atypical chemokine receptors (ACKRs) (Bachelerie, Ben-Baruch, et al., 2014; Bachelerie, Graham, et al., 2014). Although ACKRs do not form a homogenous group or cluster phylogenetically, they do share several characteristics. One of their common atypicalities is the inability to trigger the canonical G protein-mediated signaling, partly due to the lack or alterations in the DRYLAIV motif within the second intracellular loop, known to play a pivotal role in G protein coupling of classical chemokine receptors. Despite their dysfunctional G protein-mediated signaling, ACKRs modulate chemotactic responses, developmental processes and maintain and resolve immune responses. These functions are accomplished by efficient transport or internalization of chemokines into degradative compartments or their presentation on cells in order to clear abundant inflammatory chemokines, to form chemokine gradients or depots for other signaling chemokine receptors (Bachelerie, Ben-Baruch, et al., 2014; Graham, Locati, Mantovani, Rot, & Thelen, 2012; Nibbs & Graham, 2013). ACKRs are mainly expressed on epithelial cells of barrier organs, as well as lymphatic and vascular endothelial cells and to a lesser extent on circulating leucocytes. ACKRs furthermore modulate the chemokine network by directly dimerizing with canonical receptors and altering their expression or signaling properties (Kleist et al., 2016; Nibbs & Graham, 2013).

ACKRs play an essential role in tumor biology as nearly all key steps from tumor initiation to metastasis, like cancer cell proliferation, adherence to the endothelium, extravasation from blood vessels, tumor-associated angiogenesis or protection from immunological responses, are dependent on chemokine-mediated mechanisms (Keeley, Mehrad, & Strieter, 2010; Lacalle et al., 2017; Lazennec & Richmond, 2010). This chapter will outline the established and emerging roles that the atypical chemokine receptors ACKR1–4 play in cancer development, progression and metastasis, their clinical relevance, as well as the hurdles to overcome in therapeutic ACKR targeting.

2. Molecular cell biology and physiological roles of ACKRs

2.1 ACKR1

ACKR1 (formerly Duffy Antigen Receptor for Chemokines, DARC) was initially described as blood group antigen, later as receptor for the Duffy Binding Proteins (DBP) from *Plasmodium knowlesi* and *Plasmodium vivax* malaria parasites and finally as promiscuous chemokine receptor (Cutbush, Mollison, & Parkin, 1950; Horuk et al., 1993; Miller, Mason, Clyde, & McGinniss, 1976). Although ACKR1 is the oldest known chemokine receptor, it is barely recognizable as one from its primary amino acid sequence and its phylogenetic association (Neote, Mak, Kolakowski, & Schall, 1994; Nomiyama, Osada, & Yoshie, 2011). ACKR1 is the most

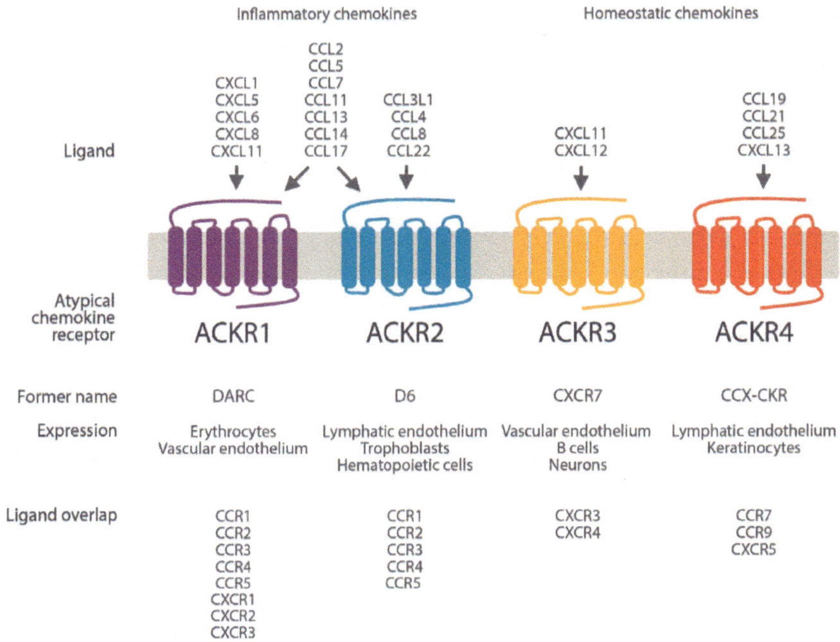

Fig. 1 ACKR expression, ligand selectivity and crosstalk with classical chemokine receptors. Atypical chemokine receptors are expressed on different types of endothelial or immune cells. ACKR1 and ACKR2 bind a broad spectrum of inflammatory chemokines that they share with CXCR1–3 and CCR1–5. ACKR3 and ACKR4 bind a more limited number of chemokines that are mainly homeostatic and are shared with CXCR3–4 and CCR7, CCR9 and CXCR5, respectively.

promiscuous chemokine receptor with over 10, primarily inflammatory, chemokine ligands from the CC and CXC chemokine families (Chaudhuri et al., 1994; Gardner, Patterson, Ashton, Stone, & Middleton, 2004; Szabo, Soo, Zlotnik, & Schall, 1995) (Fig. 1). The receptor is prominently expressed on erythrocytes and venular endothelial cells, but not on capillaries or arteries (Chaudhuri et al., 1997; Peiper et al., 1995; Thiriot et al., 2017). Depending on its cellular expression, ACKR1 serves different purposes. Unlike other atypical chemokine receptors, ACKR1 does not scavenge its ligands, but rather internalizes them in polarized cells, mediating transcytosis of intact chemokines. By doing so, ACKR1 can function as a "presenting" receptor, binding chemokines and increasing their bioavailability for other chemokine receptors in a spatiotemporally well-defined manner (Pruenster et al., 2009). Furthermore, even though ACKR1 does not seem to efficiently promote degradation of its ligands, it can compete with signaling receptors for ligand binding and reduce ligand availability in defined regions via internalization. By this mechanism,

ACKR1 was proposed to play a role in impairing chemokine-induced angiogenesis (Du et al., 2002; Horton, Yu, Zaja-Milatovic, Strieter, & Richmond, 2007).

Erythrocytic ACKR1 binds with high affinity to circulating inflammatory chemokines and can act as a "sink" or as a "buffer." Indeed, while a number of studies showed that ACKR1 modulates inflammatory responses by depleting its ligands (Darbonne et al., 1991; Dawson et al., 2000), others revealed that plasma concentrations of some ACKR1 ligands are lower in ACKR1 knockout mice as well in humans lacking ACKR1 as a natural selective plasmodium resistance, speaking in favor of a regulatory "buffer" function of the receptor (Fukuma et al., 2003; Mayr et al., 2008; Schnabel et al., 2010). It could also be shown that the lack of erythrocytic ACKR1 is associated with reduced neutrophil counts in humans (Nalls et al., 2008; Reich et al., 2009) and that ACKR1 KO mice show reduced neutrophil recruitment in inflammation models (Lee et al., 2003; Lee et al., 2006). Thus, ACKR1 can reduce the bioavailability of its ligands and prevent excessive systemic leukocyte activation but can also release bound ligands in order to avoid dramatic fluctuations in circulating chemokine concentrations. As will be described below for other ACKRs, the physiological roles of ACKR1 are further complexified by receptor heterodimerization. ACKR1 can hetero-oligomerize with CCR5 impairing the CC chemokine-induced responses, like intracellular calcium flux modulation and chemotaxis (Chakera, Seeber, John, Eidne, & Greaves, 2008).

2.2 ACKR2

ACKR2 (formerly D6 or CCBP2), was the first atypical chemokine receptor described as a "scavenger" (Fra et al., 2003). It can bind at least twelve inflammatory CC chemokines, but seems to have no affinity for homeostatic CC, nor CXC chemokines (Bonecchi et al., 2004; Bonini et al., 1997; Nibbs, Wylie, Yang, Landau, & Graham, 1997). Its main ligands include CCL2–8, CCL11–14, CCL17 and CCL22, which are agonists of the classical receptors CCR1–5 (Fig. 1). ACKR2 constitutively cycles between the cell surface and endosomal compartments. By doing so, it progressively depletes its inflammatory chemokine ligands by internalizing them and targeting them for lysosomal degradation (Galliera et al., 2004; Weber et al., 2004) and hence plays a major role in regulation and resolution of inflammatory responses.

ACKR2 is prominently expressed by lymphatic endothelial cells (LECs) of peripheral tissues and lymphoid organs (Lee et al., 2011; Nibbs et al., 2001) as well as on different leukocytes, including innate-like

B cells, but also on subsets of dendritic cells, monocytes and macrophages (Hansell et al., 2011; Kin, Crawford, Liu, Behrens, & Kearney, 2008; McKimmie et al., 2008). Furthermore, ACKR2 is expressed by various “barrier” tissues, such as skin, gut, lung, and especially placenta (Martinez de la Torre et al., 2007; Nibbs, Wylie, Yang, et al., 1997). ACKR2 knockout mice show no obvious resting phenotype. However, in line with its function as a scavenging receptor, ACKR2$^{-/-}$ mice display increased levels of inflammatory CC chemokines in different pathological contexts. This leads to the development of exacerbated inflammatory responses due to the lack of chemokine resolution and subsequent accumulation of immune cells (Jamieson et al., 2005; Martinez de la Torre et al., 2005). Recently, it was further demonstrated that ACKR2 plays a crucial role in the placenta by blocking the entry of maternal chemokines into the embryonic circulation, thus allowing embryonic chemokine gradient formation essential for directed intraembryonic cell migration (Lee et al., 2019). Although the majority of studies could convincingly identify ACKR2 as a regulatory, scavenging receptor (Bonecchi et al., 2004; Fra et al., 2003; Weber et al., 2004), other studies report signaling events like Ca^{2+} mobilization or ERK phosphorylation in ACKR2 expressing cells, contrarily to their non-expressing counterparts (Nibbs, Wylie, Pragnell, & Graham, 1997; Sjoberg et al., 2019). Moreover, ACKR2 surface expression is upregulated in response to ligand exposure, through β-arrestin-dependent activation of the cofilin signaling pathway (Borroni et al., 2013). Although these studies do not show that ACKR2 is directly generating these signaling events, it seems that ACKR2 can at least influence cellular signaling, for example, by regulating the expression levels or activity of other signaling receptors (Bonecchi et al., 2004; Hansell et al., 2011).

2.3 ACKR3

ACKR3 (CXCR7 or RDC1) is the last deorphanized chemokine receptor and binds the CXC chemokines CXCL12 and CXCL11, which also activate the classical chemokine receptors CXCR4 and CXCR3, respectively (Balabanian et al., 2005; Burns et al., 2006) (Fig. 1). ACKR3 is also a receptor for the Kaposi’s sarcoma-associated herpesvirus (HHV-8)-encoded CC chemokine vMIP-II/vCCL2 (Szpakowska et al., 2016) and was furthermore proposed to bind several non-chemokine ligands, such as the pro-angiogenic peptide adrenomedullin (Klein et al., 2014) or proenkephalin

A-derived peptides, like BAM22 (Ikeda, Kumagai, Skach, Sato, & Yanagisawa, 2013; Szpakowska, Meyrath, et al., 2018). The receptor is expressed by endothelial cells, mesenchymal cells, in diverse regions of the central nervous system and in the adrenal glands (Berahovich et al., 2014; Regard, Sato, & Coughlin, 2008; Su et al., 2002). ACKR3 seems furthermore to be expressed on B cells, but its expression profile on other leukocytes is controversial (Balabanian et al., 2005; Berahovich et al., 2010; Hartmann et al., 2008; Humpert et al., 2012; Sierro et al., 2007).

ACKR3 knockout mice die perinatally due to semilunar heart valve malformation and ventricular septal defects and show furthermore disrupted lymphangiogenesis and cardiomyocyte hyperplasia, while their hematopoiesis remains normal (Sierro et al., 2007; Yu, Crawford, Tsuchihashi, Behrens, & Srivastava, 2011). Like ACKR2, ACKR3 is primarily found intracellularly and is continuously cycling from endosomal compartments to the cell surface, a typical characteristic of a scavenging receptor (Canals et al., 2012; Luker, Steele, Mihalko, Ray, & Luker, 2010; Naumann et al., 2010). Indeed, studies done in zebrafish embryos convincingly show the role of ACKR3 scavenging function in CXCL12 gradient shaping during development (Boldajipour et al., 2008; Dona et al., 2013).

Despite its well-established function as a scavenger receptor, ACKR3 signaling potential remains debated. While numerous studies claim that ACKR3 acts solely as a scavenging receptor (Burns et al., 2006; Luker et al., 2012; Naumann et al., 2010; Wang et al., 2012), others could detect G protein-independent, β-arrestin-dependent signaling, such as ERK or AKT phosphorylation (Becker et al., 2019; Li et al., 2019; Rajagopal et al., 2010; Wang et al., 2008), and a few studies reported direct G protein-mediated signaling (Odemis et al., 2012). It is important to note that ACKR3 can interact with various other membrane receptors and alter their subcellular distribution and signaling properties, the most prominent example being the chemokine receptor CXCR4 (Hartmann et al., 2008; Levoye, Balabanian, Baleux, Bachelerie, & Lagane, 2009; Sanchez-Alcaniz et al., 2011), with which ACKR3 shares the ligand CXCL12. Overall, the signaling properties of ACKR3 seem to be multifaceted and may be cell context-dependent. It is likely that some signaling properties attributed to ACKR3 itself may rather be associated with a modified expression or signaling profile of CXCR4 by ACKR3. Mouse studies could show, for example, that both ACKR3 and CXCR4 are required for proper interneuron migration and neuronal development (Sanchez-Alcaniz et al., 2011; Wang et al., 2011). Overall, the ACKR3/CXCR4/CXCL12 trio is interdependent and is involved in a plethora of physiological processes like cell adhesion and survival, but also in various cancers (Burns et al., 2006; Sun et al., 2010; Thelen & Thelen, 2008).

2.4 ACKR4

ACKR4 (CCX-CKR or CCRL1) is expressed on keratinocytes, astrocytes, lymphatic endothelial cells and on thymic epithelial cells, but not on hematopoietic cells (Dorf, Berman, Tanabe, Heesen, & Luo, 2000; Heinzel, Benz, & Bleul, 2007; Townson & Nibbs, 2002). ACKR4 is the homeostatic complement to the inflammatory chemokine scavenger ACKR2. It binds the homeostatic chemokines CCL19, CCL21, CCL25, and with low affinity CXCL13, which are the ligands for CCR7, CCR9 and CXCR5, respectively (Fig. 1) (Gosling et al., 2000). However, mouse ACKR4 seems not to bind CXCL13 (Townson & Nibbs, 2002). Similar to ACKR3, ACKR4 is activated by the HHV-8 encoded chemokine vMIP-II/vCCL2 (Gosling et al., 2000; Szpakowska & Chevigne, 2016). Like other atypical chemokine receptors, ACKR4 does not induce any classical G protein-mediated signaling, but was shown to efficiently degrade homeostatic chemokines (Comerford, Milasta, Morrow, Milligan, & Nibbs, 2006; Heinzel et al., 2007; Townson & Nibbs, 2002; Ulvmar et al., 2014). In line with these findings, ACKR4-deficient mice show elevated levels of CCL19 and CCL21 in serum and lymph nodes (Comerford et al., 2010). These homeostatic chemokines are important mediators of naïve T-cell, B-cell and dendritic cell trafficking. Indeed, *in vivo* data show that, by scavenging these chemokines, ACKR4 regulates steady-state leukocyte migration by shaping chemokine gradients and controlling their availability in the extracellular space (Heinzel et al., 2007; Ulvmar et al., 2014). CCL19, CCL21 and CCL25 play furthermore crucial roles in thymic lymphopoiesis and deletion of ACKR4 in mice results in spontaneous autoimmunity (Bunting et al., 2013). Along these lines, ACKR4 deficient mice also develop faster and display a more severe disease onset in an EAE model attributable to increased Th17 response (Comerford et al., 2010). To our knowledge, no study could so far identify any ACKR4-mediated activation of signal transduction pathways or receptor heterodimerization.

3. The roles of ACKRs in animal tumor models

The current state of knowledge proposes a predominant tumor inhibitory effect for three of the four established ACKRs (ACKR1, ACKR2 and

ACKR4) in different animal tumor models. This is commonly believed to involve ACKR-mediated regulation of pro-tumorigenic chemokines such as CXCL8, CCL2 and CCL22. In contrast, ACKR3 that binds to CXCL11 and CXCL12 has been shown to either promote or suppress tumor growth and progression (Massara, Bonavita, Mantovani, Locati, & Bonecchi, 2016). These overall patterns derived from experimental settings are in general concordance with clinical data demonstrating that the expression of ACKR1, ACKR2 and ACKR4 in tumors correlates with a more favorable prognosis, and that of ACKR3 with an adverse outcome (as outlined in Section 4). Notably, there are exceptions to these patterns in different tumor models. As outlined below, factors contributing to the diversity of net-tumoral effects associated with individual ACKRs include the predominant cell type expressing the ACKR, the presence of other receptors for chemokines and growth factors, and the chemokine expression profile.

3.1 ACKRs in primary tumor growth and metastasis

ACKRs affect the behavior of malignant cells and different stromal cells. They can act in an autocrine manner, where the predominantly affected cell is the cell expressing the ACKR, and in a paracrine manner where ACKRs influence chemokine signaling in other cells. The best-studied example for the latter is ACKR1, which is expressed on endothelial cells and known to interact with the tetraspanin CD82/KAI1 on cancer cells. Stimulation of CD82 by ACKR1 induced tumor cell senescence by upregulation of p21 and suppressed CXCL8-mediated gap formation in endothelial cells, thus limiting tumor cell extravasation and metastasis (Bandyopadhyay et al., 2006; Khanna, Chung, Neves, Robertson, & Dong, 2014). Anti-tumoral effects of ACKR1 were also described in a prostate cancer model and linked to its scavenging capacity. Prostate tumors developing in ACKR1-deficient mice were larger compared to wild-type controls and characterized by higher vessel density and increased levels of pro-angiogenic chemokines such as CXCL2 and CXCL8 (Shen, Schuster, Stringer, Waltz, & Lentsch, 2006). In a model of pancreatic cancer, ACKR1 suppressed tumor growth and vascular invasion by counteracting CXCR2-induced activation of STAT3, a well-described promoter of tumor cell growth and invasion (Maeda et al., 2017). Furthermore, metastatic breast cancer cells engineered to express ACKR1 formed smaller tumors and less lung metastases in mouse xenografts (Wang et al., 2006).

ACKR2 also limits tumorigenesis as demonstrated, e.g., in a tumor model of Kaposi sarcoma. Downregulation of ACKR2 in Kaposi sarcoma cells gave rise to larger tumors due to increased CCL2-mediated recruitment of monocytes and concomitant differentiation to pro-angiogenic, tumor-stimulatory M2-like macrophages in the tumor microenvironment (Savino et al., 2014). The enhanced expression of ACKR4 in cancer cells has also been found to limit tumor growth and metastasis by inhibiting tumor cell migration and invasion, and by antagonizing the action of chemokine receptors including CCR7, CCR9, and CXCR5 that can enhance malignancy (Feng et al., 2009; Shi et al., 2015; Zhu et al., 2014).

ACKR3 instead was demonstrated to promote proliferation, migration and invasion of different types of cancer cells resulting in enhanced tumor growth and metastasis (Li et al., 2015; Luker et al., 2012; Miao et al., 2007; Puddinu et al., 2017; Werner et al., 2017). In an ovarian cancer model, estrogen was shown to induce the expression of ACKR3 that favored tumor cell migration, invasion and epithelial-mesenchymal transition (EMT), possibly through CXCL11 (Benhadjeba, Edjekouane, Sauve, Carmona, & Tremblay, 2018). However, CXCL12-stimulated EMT in ovarian cancer cells seemed to rely rather on CXCR4 than ACKR3, suggesting a context-dependent contribution of ACKR3 to EMT-signaling (Zheng et al., 2019). An important question is how ACKR3 confers pro-tumorigenic effects, while the other ACKRs are predominantly reported for their tumor inhibitory action. One proposed mechanism links ACKR3 to other pro-tumorigenic signaling receptors such as the estrogen receptor and epidermal growth factor receptors (EGFR) (Hao et al., 2018; Salazar et al., 2014). Several other studies suggest that ACKR3 increases the availability of CXCR4 on the cell surface by different ways, e.g., scavenging of CXCL12 or prevention of CXCR4 internalization and degradation (Abe et al., 2014; Coggins et al., 2014). Furthermore, heterodimerization of ACKR3 with CXCR4 that modulates the signaling properties and interactions of CXCR4 with intracellular effectors was introduced as another potential mechanism (Levoye et al., 2009). Thus, the pro-tumorigenic activity of ACKR3 is presumably linked to its interplay with membrane receptors that contribute significantly to tumor growth and metastasis (Walenkamp, Lapa, Herrmann, & Wester, 2017). However, the overall picture appears more complex as other data showed that the co-expression of CXCR4 and ACKR3 inhibited CXCL12-stimulated pro-metastatic signaling (Hernandez, Magalhaes, Coniglio, Condeelis, & Segall, 2011). These examples illustrate the dual effects of ACKR3 in cancer growth and progression.

3.2 ACKRs in angiogenesis

A role of ACKRs in angiogenesis includes the sequestration of pro-angiogenic chemokines as well as the inhibition of other regulators of vessel growth and maturation (Savino et al., 2014; Shen et al., 2006) ACKR1, ACKR2 and ACKR3 are expressed on endothelial cells and modulate their behavior and local microenvironment by regulating the bioavailability and transcytosis of chemokines through the endothelial layer. ACKR1$^{-/-}$ mice display increased levels of the angiogenic chemokines CXCL8 and CXCL2 affecting the vascularity of tumors (Shen et al., 2006). Conversely, xenograft tumors established from the injection of ACKR1-overexpressing NSCLC cells are characterized by lower blood vessel density that likely contributed to the enhanced necrosis observed (Addison, Belperio, Burdick, & Strieter, 2004). Overexpression of ACKR1 by breast cancer cells prevented the accumulation of tumor cell-secreted CXCL8 and CCL2, thus inhibiting microvascular density and the composition of the extracellular matrix in the tumors formed (Wang et al., 2006). A similar impact of ACKR1 on microvascular density and tumor growth has been described in an ovarian cancer model (Zhu, Jiang, & Wang, 2017). Furthermore, transgenic expression of ACKR1 in endothelial cells inhibits tumor angiogenesis and growth in a melanoma tumor model (Horton et al., 2007). For ACKR3 opposing roles in tumor angiogenesis have been reported. On one hand, ACKR3 can affect angiogenesis through regulation of CXCL12 levels, and sequestration of CXCL12 limits endothelial precursor recruitment, blood vessel formation and tumor growth (Orimo et al., 2005; Stacer et al., 2016). In contrast, overexpression of ACKR3 can also promote angiogenesis as shown for ACKR3-expressing prostate cancer cells that secrete enhanced levels of CXCL8 and VEGF (Wang et al., 2008). Thus, also with regard to angiogenesis ACKR3 seems to display diverse and context-dependent effects.

3.3 ACKRs in inflammation and tumor immunity

Inflammation is a hallmark of cancer, and (re)activation of the host immunity has become an important approach in cancer therapy. ACKRs, mainly ACKR2 and ACKR3, are expressed by a variety of immune cells and act as critical regulators of immune cell behavior and contribute to shaping the tumor microenvironment. For example, ACKR2 seems to control the accumulation of a specific subpopulation of anti-tumoral NK cells in tumors. Knocking down ACKR2 in mice increased the intra-tumoral numbers of a NK-cell subset with high expression of CCR2. These NK cells were recruited to CCL2-expressing tumor cells in the metastatic niche and thereby restricted metastasis. Of note, the number of other CCR2-positive immune cells including monocytes was not affected in ACKR2-deficient mice suggesting an immune cell subset-specific role for ACKR2 in mice (Hansell et al., 2018).

In addition to effects of immune cell-expressed ACKRs, cancer- and other stroma cell-expressed ACKRs can also affect innate and adaptive immune responses and inflammation. ACKR2 is known to negatively control the recruitment of pro-inflammatory cells such as monocytes and neutrophils to sites of inflammation by scavenging a broad range of pro-inflammatory chemokines including CCL2 and CCL3 (Rot et al., 2013; Savino et al., 2014). Consequently, genetic deletion of ACKR2 was shown to enhance tumor-stimulatory processes such as leukocyte mobilization, inflammation and angiogenesis. ACKRs do not only control inflammation, they are also regulated by pro-inflammatory signaling. Expression of ACKR3 is induced upon several stimuli including pro-inflammatory signals from, e.g., rheumatoid arthritis, inflammatory bowel disease, virus and cancer (Sanchez-Martin, Sanchez-Mateos, & Cabanas, 2013). On the contrary, psoriasis-associated inflammation repressed the expression of ACKR2 by induction of mir-146b and miR-10b (Shams et al., 2018). Also in cancer, ACKR2 seems to be a target of miRNA as demonstrated for the miR-146a-mediated down-regulation of ACKR2 in anaplastic thyroid carcinomas (Pacifico et al., 2017).

Thus, ACKRs play a significant role in inflammation and immunity due to their expression on a variety of immune cells, and their function as scavengers of immune-regulatory chemokines.

3.4 Dual, tumor context-specific effects of ACKRs

As outlined above, ACKR1, ACKR2 and ACKR4 have mostly been associated with anti-tumoral effects whereas ACKR3 is more linked to tumor-promoting activities. However, there are also findings which indicate that this dichotomization is over-simplified, not only for ACKR3, as discussed in previous paragraphs, but also for ACKR1, 2 and 4.

For example, overexpression of ACKR1 in lung cancer cells reduced metastasis while primary tumors grew significantly larger (Addison et al., 2004; Wang et al., 2006). Similarly, more recently published work proposed that ACKR2 can support tumor growth and progression (Hansell et al., 2018; Sjoberg et al., 2019). Using the PyMT, transgenic breast cancer mouse model and metastasis models with B16-F10 and LLC cancer cells, Hansell et al. found impaired metastatic potential of tumor cells in mice with ACKR2 knockdown. This effect was attributed to the enhanced recruitment of CCR2-expressing, tumor-killing NK cells to the CCL2-secreting tumors in ACKR2-depleted mice (Hansell et al., 2018). ACKR2 is also present on stromal cells, including cancer-associated fibroblasts, and was recently described to mediate the effects of the chemokine CXCL14 that, in this setting, promotes tumor growth and metastasis (Sjoberg et al., 2019). The context-dependent action of ACKR2 is further highlighted by the finding that primary tumor growth of NeuT-driven breast cancer cells was enhanced in ACKR2-knockout mice but metastasis was suppressed (Massara et al., 2018). Harata-Lee et al. also found that the size of orthotopically growing 4T1.2 tumors was reduced when ACKR4 was overexpressed in these cells but the rate of spontaneous metastasis was enhanced, suggesting a pro-metastatic effect of ACKR4 under these conditions (Harata-Lee et al., 2014).

Taken together, ACKRs play a pivotal role in a variety of physiological processes that become perturbed in cancer. To date, there is compelling evidence for ACKRs to inhibit tumor growth, angiogenesis and metastasis by regulating the availability of different pro-tumorigenic chemokines. However, ACKRs can also support and stimulate the malignancy of tumors by interacting with different types of receptors, by exposition to a tumor-specific chemokine milieu or by their tumor-induced, enhanced expression in cell types that under physiological conditions express no or low levels of ACKRs. It is therefore clear that a more profound level of knowledge is needed for successful therapeutic targeting of ACKRs in cancer.

4. ACKR expression in patient samples and correlation with clinical outcome

In this section we discuss how the individual expression of ACKRs in tumor tissue, or their co-expression, is linked to patient characteristics, survival and predicting therapy response. The clinical relevance of ACKRs has been explored in various studies of different human cancers. The expression of ACKR1, ACKR2 and ACKR4 correlates with improved survival and shows an inverse correlation with poor prognosis-associated clinicopathological parameters, supporting the tumor inhibitory effects discussed above. ACKR3 has instead been shown to correlate with an impaired prognosis and protein levels associate positively with advanced tumor stage and presence of lymph node metastasis (see summary in Table 1).

Table 1 Protein expression of ACKRs in human tumors and correlation with clinicopathological parameters and survival.

Atypical chemokine receptor	Tumor type	Correlation with prognosis	Association to clinicopathological parameters	Cell type expression	Expression in cancer as compared to normal tissue	Publication
ACKR1	PDAC	Improved RFS and OS	Tumor size (−), histological grade (−), vascular invasion (−)	Tumor cells	Upregulated	Maeda et al. (2017)
	Colorectal	Not investigated	Differentiation (+), MV density (−), TNM stage (−), LN metastasis (−)	N.D.	Downregulated	Zhou et al. (2015)
	Breast	Improved OS	MV density (−), ER status (+), Metastasis (−)	N.D.	N.D.	Wang et al. (2006)
	Breast	Improved RFS and OS	TNM stage (−), LN metastasis (−)	Mainly tumor cells	No difference	Zeng et al. (2011)
	Laryngeal	Improved OS	MVD (−), Local recurrence (−), TNM stage (−)	N.D.	Upregulated	Sun, Wang, Zhu, Huang, and Ji (2011)
	Cervical	No impact on RFS or OS	LN metastasis (−)	N.D.	Downregulated	Hou et al. (2013)
	Gastric	Improved OS	Histological grade (−), Macroscopic type, TNM stage(−)	N.D.	No difference	Zhu et al. (2013)

ACKR2	Breast	Improved DFS	LN metastasis (−)	Tumor cells	N.D.	Wu et al. (2008)
	Breast	Improved RFS and OS	TNM stage (−), LN metastasis (−)	Mainly tumor cells	No difference	Zeng et al. (2011)
	Cervical	Improved RFS and OS	Tumor size (−), Recurrence (−)	N.D.	Downregulated	Hou et al. (2013)
	Gastric	Improved OS	Histological grade (−), Macroscopic type, TNM stage (−)	N.D.	No difference	Zhu et al. (2013)
ACKR3	Breast	Worse OS	HER2 status (+), LN metastasis (+)	N.D.	Upregulated	Wu, Qian, Chen, and Ding (2015)
	Colorectal	Worse PFS and OS	TNM stage (+), LN metastasis (+), distant metastasis (+)	N.D.	Upregulated	Yang et al. (2015)
	Cervix	Worse DFS, DSS	EGFR status (+), Tumor size (+), TNM stage (+)	N.D.	N.D.	Schrevel et al. (2012)
	Glioma	Worse survival (IDH1 mutant patients)	WHO grade (+), IDH1 mutation (−)	Endothelial cells and tumor cells	N.D.	Birner, Tchorbanov, Natchev, Tuettenberg, and Guentchev (2015)
	Glioma	No impact on OS	IDH1 mutation (−)	Endothelial cells and tumor cells	N.D.	Bianco et al. (2015)

Continued

Table 1 Protein expression of ACKRs in human tumors and correlation with clinicopathological parameters and survival.—Cont'd

Atypical chemokine receptor	Tumor type	Correlation with prognosis	Association to clinicopathological parameters	Cell type expression	Expression in cancer as compared to normal tissue	Publication
ACKR4	Breast	Improved RFS and OS	TNM stage (−), LN metastasis (−)	Mainly tumor cells	No difference	Zeng et al. (2011)
	Gastric	No difference	Age (+), Macroscopic type, TNM stage (−)	N.D.	No difference	Zhu et al. (2013)
	Cervical	Improved OS	Tumor size (−), FIGO stage (−), LN metastasis (−)	N.D.	Downregulated	Hou et al. (2013)
	Breast	Improved OS	LN metastasis (−)	N.D.	N.D.	Feng, Ou, Wu, Shen, and Shao (2009)
	Liver	Improved RFS and OS	Vascular invasion (−), Tumor differentiation (+)	N.D.	Downregulated	Shi et al. (2015)
	Colorectal	Improved PFS and OS	TNM stage (−), LN metastasis (−)	N.D.	Downregulated	Zhu et al. (2014)

PDAC = Pancreatic ductal adenocarcinoma, OS = Overall survival, RFS = recurrence free survival, PFS = Progression free survival, DFS = Disease-free survival, DSS = Disease specific survival, TNM = Tumor node metastasis, LN = lymph node, MV = micro vessel, (−) = Negative association, (+) = positive association, N.D. = Not determined.

4.1 ACKR1 and ACKR2

Clinical relevance of ACKR1 expression has been obtained from numerous studies of different tumor types including laryngeal squamous cell carcinoma (LSCC), pancreas, breast, cervical, thyroid and colorectal cancer (CRC) (Hou et al., 2013; Latini et al., 2013; Maeda et al., 2017; Sun et al., 2011; Wang et al., 2006; Zeng et al., 2011; Zhou et al., 2015). Less attention has been paid to the importance of ACKR2 expression in clinical samples, but expression has been correlated to prognosis in studies of breast, cervical and gastric cancer (Hou et al., 2013; Wu et al., 2008; Yu, Wang, Yang, Zeng, & Shao, 2015; Zeng et al., 2011; Zhu et al., 2013). Results from published reports mainly suggest a protective function and association with improved patient outcome for the two receptors.

In pancreatic ductal adenocarcinoma (PDAC) ACKR1 was found to be expressed in the tumor cells, whereas protein expression was weak in normal pancreas and in the peritumoral area. Patients with low protein levels of ACKR1 exhibited tumors with poor differentiation, larger size and increased vascular invasion. Uni- and multivariable analyses demonstrated a significant correlation between high ACKR1 levels and improved relapse-free (RFS) and overall survival (OS) (Maeda et al., 2017). On the contrary, in a study comparing ACKR1 expression in 90 CRC patients with 64 paired unaffected tissues, mRNA and protein levels of the receptor was instead found to be down regulated in tumors compared to non-cancerous colon. However, similar to results from PDAC, associations to clinicopathological features identified ACKR1 to associate positively with tumor differentiation, and negatively with lymph node (LN) metastasis and tumor node metastasis (TNM) stage. In addition, tumors with high ACKR1 had significantly lower microvascular density (Zhou et al., 2015). In concordance, immunohistochemistry (IHC) analysis of ACKR1 levels in a cohort of 75 breast cancer patients also revealed a negative correlation with microvascular density, LN metastasis and estrogen receptor (ER) status. Expression of the receptor was also associated with better patient survival (Wang et al., 2006). In a larger cohort, including 463 patients with invasive breast cancer, protein expression of ACKR1 and ACKR2 was correlated positively to RFS (Yu et al., 2015). In agreement, in a study by Wu et al., both mRNA and protein levels of ACKR2 were found to predict better breast cancer-free survival and mechanistic data obtained from *in vivo* experiments, showed an inhibitory effect on tumor invasion and metastasis formation (Wu et al., 2008). As opposed to expression analysis in breast tumors, where ACKR2 is downregulated, ACKR2 protein was revealed to be highly expressed in vascular tumors, including angiosarcomas and Kaposi's sarcomas (Nibbs et al., 2001). Numerous studies have also explored the survival effects of combined expression of ACKR1, ACKR2 and ACKR4 in cancer, which is further discussed below.

The importance of genetic variations in ACKRs have been explored in breast cancer (Li, Yang, Shao, & Yu, 2018; Yang et al., 2013; Yu et al., 2015). Evidence from two reports demonstrated that allele variations in ACKR1 and ACKR2 contribute to cancer survival and clinical features. Two single nucleotide polymorphisms (SNPs) in ACKR1 (rs12075) and ACKR2 (rs2228468), respectively, were correlated to increased LN metastasis and worse RFS in breast cancer patients (Yang et al., 2013; Yu et al., 2015). A follow-up study by the same authors further analyzed these two non-synonymous polymorphisms in 806 breast cancer patients and correlated the genotypes to cancer recurrence after adjustment to LN status. Only the SNP for ACKR1, and not for ACKR2, showed clinical impact and patients with a GG genotype in rs12075 had a significantly worse outcome, as compared to patients with the AG/AA genotype. Moreover, subgroup-specific analysis revealed an increased risk of recurrence for the GG genotype only in triple negative patients, as compared to the other molecular subtypes (Li et al., 2018).

All prognostic studies discussed above demonstrate a survival benefit of having ACKR1 or ACKR2 present in the tumor tissue. However, these studies did not, in general, consider cell type-specific expression of the receptors. Interestingly, the impact of ACKR1 expression in erythrocytes on prostate cancer risk has been investigated in a study by Elson et al. However, authors found no link between a genetic loss of ACKR1 in red blood cells and risk of developing prostate cancer or a more aggressive disease (Elson et al., 2011). Recent work on the orphan chemokine CXCL14 derived from cancer-associated fibroblasts identified an autocrine CXCL14/ACKR2 pathway stimulating breast cancer invasion, metastasis and survival (Sjoberg et al., 2019). Concordant with these experimental findings, analyses of the TCGA gene expression dataset of breast cancer revealed that CXCL14 only predicts worse OS in patients with high mRNA expression of ACKR2 (Sjoberg et al., 2019).

4.2 ACKR3

Altered expression of ACKR3 has been demonstrated in several cancers, including prostate, kidney, liver, colon, cervix, brain, lung and breast cancer

(Freitas et al., 2014; Sanchez-Martin et al., 2013). In comparison with results from prognostic studies of ACKR1 and ACKR2, survival analyses for ACKR3 expression has instead shown a correlation to worse outcome (Hassan et al., 2009; Ierano et al., 2014; Schrevel et al., 2012; Wu et al., 2015; Wu, Tang, Sun, & Sun, 2016; Yang et al., 2015; Yu et al., 2017).

In patients with CRC, ACKR3 mRNA and protein expression was found to be elevated as compared to normal colon. Protein expression of the receptor was associated with higher TNM stage and distant metastasis and correlated to worse progression-free survival (PFS) and OS. Multivariable cox-regression analysis adjusting for known clinicopathological factors, identified ACKR3 to independently predict worse CRC outcome (Yang et al., 2015). In agreement, IHC staining of 103 formalin-fixed paraffin embedded specimen of cervical cancer showed that high ACKR3 levels were independently correlated to worse disease-free and disease-specific survival and significantly positively associated with larger tumor size and LN metastasis (Schrevel et al., 2012). In lung adenocarcinoma, mRNA expression of ACKR3 was identified as a poor-prognostic marker enhancing tumor growth and TGFβ mediated EMT (Wu et al., 2016). In concordance, ACKR3 together with CXCR4 predicted worse prognosis in renal cell carcinoma patients instead by activating the mTOR pathway (D'Alterio et al., 2010; Ierano et al., 2014). In congruence, several breast cancer studies show an increase in both mRNA and protein expression of ACKR3 in tumor tissue as compared to non-cancerous breast (Li et al., 2015; Miao et al., 2007; Wani et al., 2014). High receptor expression has been shown to predict poor patient prognosis and associate positively with lymph node metastasis (Hassan et al., 2009; Li et al., 2015; Wani et al., 2014; Wu et al., 2015; Yu et al., 2017). The cell type-specific expression of ACKR3 in tumors has been investigated in several studies. ACKR3 was shown to be expressed by the tumor vasculature in glioma patients. Survival analysis revealed that endothelial expression of the receptor significantly correlated with improved outcome in IDH wild-type tumors and worse outcome in IDH mutated tumors (Birner et al., 2015; Calatozzolo et al., 2011). The expression pattern of ACKR3 has also been shown to change with tumor grade. In grade II gliomas, ACKR3 is expressed by the tumor cells, whereas grade III and glioblastoma (grade IV) also exhibited enhanced expression in the endothelial cells (Bianco et al., 2015). Similarly as for glioma, the expression of ACKR3 in osteosarcoma, breast, lung, colon and renal cancer was localized to the tumor endothelial cells and no expression of the receptor was observed in normal vasculature (Goguet-Surmenian et al., 2013; Guillemot et al., 2012; Maishi et al., 2012; Miao et al., 2007).

4.3 ACKR4

Evidence from tumor models of the involvement of ACKR4 in tumor progression is conflicting, as discussed above. ACKR4 expression has been shown to enhance the metastatic potential of breast cancer by modulating EMT and tumor cell invasion (Harata-Lee et al., 2014). However, clinical relevance of these mechanistic data has not been proven. On the contrary, high ACKR4 expression in tumor tissue of breast, liver and colon showed a significant correlation with improved patient prognosis (Feng et al., 2009; Shi et al., 2015; Zhu et al., 2014).

High ACKR4 expression was associated with lower incidence of LN metastasis in 98 human breast cancers, evaluated by IHC. Furthermore, patients expressing ACKR4 survived longer, as compared to patients with low ACKR4-expressing tumors. Sub-group analysis revealed that only patients with LN metastasis exhibited a survival benefit, as compared to patients with no metastasis in the LN. Moreover, ACKR4 was identified as an independent prognostic marker for disease-free survival (Feng et al., 2009). In two independent cohorts of hepatocellular carcinoma, ACKR4 expression was evaluated and correlated with clinical significance. Both mRNA and protein expression was downregulated in liver cancer as compared to normal liver. In both the discovery and validation cohorts, ACKR4 deficiency was associated with advanced tumor stage and was shown to be an independent predictor for worse survival and increased risk for recurrence (Shi et al., 2015). In a study by Zhu et al., the expression levels of ACKR4 was analyzed in 136 paired samples of CRC an normal tissue. As in other tumor types, data showed a decrease in ACKR4 protein levels in CRC, as compared to normal tissue. In addition, ACKR4 protein expression was negatively correlated to LN metastasis and positively correlated to patient survival. The receptor was also identified as an independent factor for PFS in multivariable analysis (Zhu et al., 2014).

4.4 Co-expression of ACKR1, ACKR2 and ACKR4

The clinical relevance of ACKR1, ACKR2 and ACKR4 protein expression, and their co-expression have further been explored in breast cancer, cervical squamous cell carcinoma (CSCC) and gastric cancer (Hou et al., 2013; Zeng et al., 2011; Zhu et al., 2013). In breast cancer, co-expression of all the receptors was decreased in invasive breast cancer as compared to non-invasive cancer and normal breast. Furthermore, combined high expression was associated with LN negative tumors, lower TNM stage, and predicted a favorable prognosis both in uni- and multivariable analysis. Individual expression of the receptors also correlated with increased RFS and OS in univariable analysis, but only to RFS in multivariable analysis (Zeng et al., 2011). In 227 cases of cervical squamous cell carcinoma, expression of ACKR1, ACKR2 and ACKR4 was analyzed for associations with clinicopathological characteristics and survival time. Only ACKR1 and ACKR4 and co-expression of all receptors significantly correlated with enhanced OS, whereas no significant difference in survival was observed between ACKR2-negative and -positive tumors (Hou et al., 2013). Lastly, expression of the three receptors individually, and their co-expression, was correlated with better outcome in patients with gastric cancer. Co-expression of ACKR1 ACKR2 and ACKR4 was also identified as independent prognostic predictor of enhanced OS (Zhu et al., 2013).

4.5 ACKRs predicting therapy response

The relevance of ACKRs in predicting response to treatment has not been extensively studied. A few recent reports have shown a link between ACKR3 expression and therapy-resistant disease, but no studies have yet explored the importance of ACKR1, ACKR2 and ACKR4 as response-predicting markers. In a small NSCLC patient cohort, ACKR3 was overexpressed in patients harboring EGFR mutated tumors and the increased expression activated MAPK signaling, leading to resistance to EGFR inhibitors (Becker et al., 2019). ACKR3 might also serve as a biomarker of therapy resistance in prostate cancer patients. Li et al. showed that castration resistant prostate cancer, with resistance to the androgen receptor antagonist enzalutamide, showed enhanced levels of ACKR3 and the receptor was highly expressed in the tumor endothelial cells. The mechanism behind the resistance also involved activation of MAPK signaling and targeting of the receptor may help overcoming the development of resistant tumors (Li et al., 2019). The role of ACKR3 as a predictive marker for chemotherapy resistance was explored in esophageal squamous cell carcinoma (ESCC). ACKR3 protein levels were analyzed in 45 paired ESCC and adjacent non-cancerous tissue. The levels of the receptor were increased in tumor tissue as compared to non-tumor tissue and further enhanced in patients resistant to cisplatin treatment. Moreover, patients receiving cisplatin post-operatively were analyzed for the expression of ACKR3. Patients with tumors high in ACKR3 expression had worse PFS and OS, as compared to patients with low ACKR3 expression in tumor tissue (Qiao et al., 2018).

5. ACKR-specific modulators for cancer therapy

Considering the complexity of the interactions in which ACKRs participate and their numerous context-dependent functions, the therapeutic targeting of this family of receptors is challenging. While so far no therapeutics directed against ACKRs have been accepted or tested in clinical trials, ACKR-specific molecules or antibodies have been described and used to investigate the pharmacology of ACKRs and their roles in different physio-pathological processes. The vast majority of these molecules pertain to ACKR3.

5.1 Small-molecule modulators

Several small molecules are known to interact with ACKR3. CCX771 is one of the most potent ACKR3 modulators inducing β-arrestin recruitment to the receptor (Zabel et al., 2009) and was reported to inhibit tumor growth, lung metastasis and tumor angiogenesis *in vivo* (Yamada et al., 2015). In prostate cancer models, CCX771 was also shown to promote significant tumor suppression in combination with enzalutamide treatment, which was proposed as partly due to a reduction in pro-angiogenic signaling and in the formation of large blood vessels in tumors (Luo et al., 2018). Other analogues have been developed with various pharmacological profiles, including the partial agonist CCX777 (Gustavsson et al., 2017) or CCX733, which has been reported to act as ACKR3 antagonist (Hartmann et al., 2008). Together with VUF11207 and VUF11403, small molecule agonists designed on a similar scaffold (Wijtmans et al., 2012), these compounds are valuable tools for investigating ACKR3 biology and may lead to development of novel cancer therapeutics.

The bicyclam AMD3100 (known as Plerixafor or Mozobil®) approved by the US Food and Drug Administration (FDA) in 2008 for hematopoietic stem cell mobilization and autologous transplantation in patients with non-Hodgkin's lymphoma and multiple myeloma (Kalatskaya et al., 2009) has also been shown to act on ACKR3. However, AMD3100 shows low affinity toward ACKR3, its main target being CXCR4 (Donzella et al., 1998; Schols, Este, Henson, & De Clercq, 1997).

5.2 Peptide-derived modulators

Recently, cell-permeable macrocyclic peptide-peptoid hybrid nanomolar ACKR3 modulators have been developed (Boehm et al., 2017). These compounds may provide advantages in terms of potency, selectivity, and reduced off-target activity compared to small-molecule modulators. However, their oral absorption and metabolic stability still need to be improved. It was also shown that the cyclic peptidomimetic TC14012, which is suggested to provide therapeutic advantage by inhibiting the CXCR4-CXCL12 axis in chronic lymphocytic leukemia (CLL) (Burger et al., 2005; Juarez, Bradstock, Gottlieb, & Bendall, 2003) can interact with ACKR3 (Gravel et al., 2010). However, its potential to inhibit ACKR3-dependent tumorigenesis has not been evaluated.

5.3 Ligand-based modulators

Besides these approaches, chemokine-derived modulators such as the CXCL11/CXCL12 chimeric chemokine or peptides derived from the N termini of ACKR3-related chemokines were also developed and optimized for enhanced ACKR3 potency and selectivity (Ameti et al., 2018; Chevigne, Fievez, Schmit, & Deroo, 2011; Fievez et al., 2018; Szpakowska, Nevins, et al., 2018). Recently, LIH383, an octapeptide derived from an endogenous ACKR3 ligand, was also reported as a sub-nanomolar agonist of the receptor. These ligand-derived molecules are powerful pharmacological or imaging tools with great potential as chemical biology tools and as drug lead compounds.

5.4 Monoclonal antibodies and antibody fragments

Several ACKR3-specific monoclonal antibodies are routinely used for research, including clones 8F11 and 11G8, which are able to inhibit the binding of CXCL11 and CXCL12 as well as CXCL11- and CXCL12-induced β-arrestin association with the receptor (Zabel et al., 2009). It has been shown that ^{89}Zr-labeled 11G8 is able to detect ACKR3 in NOG mice xenografted with human breast, lung and oesophageal cancers, suggesting that ACKR3 is a viable candidate diagnostic marker (Behnam Azad et al., 2016). Recently an anti-ACKR3 single chain antibody (X7Ab) with a human immunoglobulin G1 (IgG1) FC sequence based on the FDA-approved Rituximab FC domain has been described (Salazar et al., 2018). It binds to the same site on the receptor as CXCL12 and inhibits CXCL12-mediated receptor activation. This engineered X7Ab is able to engage an anti-tumor immune response through FC-driven antibody-dependent cell cytotoxicity (ADCC) and antibody-dependent cellular phagocytosis (ADCP). It was shown to mediate PBMC-driven and NK-driven ADCC killing of glioblastoma U343, U251X7, and GL261 cells and in combination with the alkylating agent temozolomide (TMZ) significantly slowed cancer progression in glioblastoma mouse models. X7Ab-TMZ combination therapy may therefore allow to reduce the dose of TMZ and lower adverse effects associated with this therapy.

ACKR3-specific antibody fragments derived from the camelid heavy-chain-only antibodies, called nanobodies, with anti-tumoral effect have also been described. A nanobody selected to block ACKR3 interaction with CXCL12 reduced tumor growth in head and neck cancer cells via its anti-angiogenic properties (Maussang et al., 2013). It has also been suggested that associating the anti-tumoral effect of ACKR3 antagonists with CXCR4 blockade reducing CXCL12-mediated migration could offer a therapeutic advantage in cancer therapy (Adlere et al., 2019).

Regarding other ACKRs, despite the increasing number of studies showing their implication in cancer development and progression, few antibodies and no small molecules recognizing ACKR1, ACKR2 or ACKR4 have been reported.

6. Future perspectives

Considerable progress has been made over the last decade regarding the molecular biology of ACKRs and their implication in cancer. The current state of knowledge proposes complex tumor regulatory effects of the different ACKRs. Clinical relevance of experimental findings is also suggested by results from emerging correlative analyses of tumor tissue collections. Finally, efforts have been initiated to develop ACKR-modulatory small molecules and antibodies paving the road for therapeutic intervention. The following paragraphs discuss and suggest some areas that appear important for continued development toward a better understanding of the implication and clinical exploitation of ACKRs in cancer.

Considering the multiple effects of ACKRs in different processes that are the hallmarks of cancer (Fig. 2), forthcoming studies on the roles of ACKRs in tumor biology should accommodate the concept that the net effects of ACKRs on different aspects of tumor biology (e.g., growth, metastasis, immune surveillance) can vary depending on the cell type in which the individual ACKR is expressed. ACKRs have indeed been shown to be present

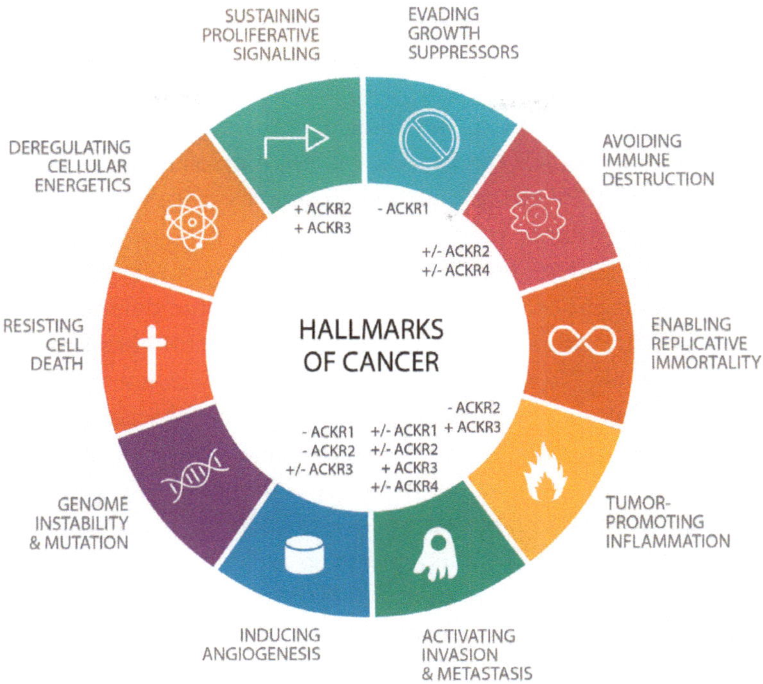

Fig. 2 ACKRs and the hallmarks of cancer. The four atypical chemokine receptors have been reported to modulate several processes necessary for cancer development and progression. The pro-tumoral and anti-tumoral effects of the individual ACKRs are presented within the framework of cancer hallmarks and indicated with + or −, respectively. *Adapted from Hanahan, D., & Weinberg, R. A. (2011). Hallmarks of cancer: The next generation.* Cell, *144(5), 646–674. https://doi.org/10.1016/j.cell.2011.02.013.*

in both tumor cells and cells within the tumor microenvironment where they may have either similar or opposing effects. Moreover, there are still multiple aspects of tumor biology where the roles of ACKRs have not been systematically analyzed. This includes the involvement of ACKRs, e.g., in cancer stem cell support, contributions to pre-metastatic niches and a potential impact of ACKRs on hallmarks they are not yet assigned to (Fig. 2). Studies in these areas will possibly suggest novel attractive pathways for drug targeting.

A key issue regarding the molecular cell biology of ACKRs is the continued delineation of effects on tumors that can be attributed to the scavenging function of ACKRs, to their direct interaction and modulation of the activity of classical chemokine receptors or their ability to induce G protein-independent signaling pathways.

Earlier studies relying on the use of knockout mice are therefore expected to be followed by more stringent and relevant analyses in mouse models using cell- or compartment-specific depletion of the different ACKRs. Similarly, studies in cell lines overexpressing ACKRs or using downregulation of ACKRs expression will be complemented and gradually replaced with CRISPR-edited cell lines or primary cells, which will represent more accurate tumor models.

Forthcoming analyses of ACKR-profiling of clinical samples should also consider the concept of cell type-specific effects of the different ACKRs. New methods are emerging for multiplexed tissue profiling (Datar et al., 2019; Goltsev et al., 2018; Savas et al., 2018). These will allow studies where a panel of cell-type-specific antibodies can be combined with ACKR antibodies. Together with digital image analyses, these studies will allow refined tissue profiling, which will be able to describe biologically more informative outcome associations, e.g., by cell type-specific scoring in the same sections of multiple ACKRs. Such multiplex studies will also allow to investigate where cell type-defined expression of different ACKRs can be related to tumor cell features such as proliferation or sternness, and immune cell composition. These approaches will thus not only have value from a candidate biomarker perspective but can also be considered as an approach for discovery of novel clinically relevant ACKR-related biology. Notably, a key resource to be developed, for these opportunities to be explored, are better mAbs for IHC/IF applications.

The development of successful drugs targeting chemokine receptors is extremely challenging (Schall & Proudfoot, 2011) but the recent reports on novel ACKR-modulatory compounds are promising and encouraging. Although so far mainly targeting ACKR3, drugs and antibodies targeting ACKR1, ACKR2 and ACKR4 will start to emerge. Recent data and novel methods in drug design, nano- or personalized medicine also point toward the possibility to generate either multi-functional or combined therapies including one ACKR-targeting component.

Finally, the ACKR familly may be enlarged in the coming years to welcome additional receptors. Indeed the names of ACKR5 and ACKR6 have already been reserved for the receptors CCRL2 (Leick et al., 2010) and PITPNM3 (Chen et al., 2011), respectively (Bachelerie et al., 2015). Even though they await functional validation with regard to chemokine-binding and atypical signaling properties, both receptors have already been proposed to have implications in different cancer onsets (Catusse et al., 2010; Lin et al., 2016; Liu et al., 2019; Shin, Zabel, & Pachynski, 2018). Further studies will reveal whether these receptors share common functional properties with the other established atypical chemokine receptors, providing new insights into chemokine regulation in tumor development but also further complexifying the global picture of atypical chemokine receptors in cancer.

In summary, ACKR studies are in a dynamic phase. Their functional complexity, differences but also multilevel involvment should not be neglected and might at first glance reduce enthusiasm for continued studies toward ACKR-targeting drugs. However, strong phenotypic findings remain from experimental models where different ACKRs have been manipulated. Optimism can thus be maintained regarding possibilities to reach the overall goal of targeting this fascinating class of receptors for cancer therapeutic purposes.

Acknowledgments

This study was supported by the Luxembourg Institute of Health (LIH) MESR (Grants 20160116 and 20170113), Luxembourg National Research Fund (INTER/FWO "Nanokine"—Grant 15/10358798), F.R.S.-FNRS-Télévie (Grants 7454719, 7459319, 7456814 and 7461515). M. Meyrath is Luxembourg National Research Fund PhD fellow (Grants AFR-3004509 and PRIDE-11012546 "NextImmune"). Studies in the AÖ group are supported from grants from Swedish Cancer Society, The Swedish Research Council, "Radiumhemmets forskningsfonder," EU ITN program and Stockholm City Council. The authors wish to thank Joanna Muz for her assistance in figure design.

References

Abe, P., Mueller, W., Schutz, D., MacKay, F., Thelen, M., Zhang, P., et al. (2014). CXCR7 prevents excessive CXCL12-mediated downregulation of CXCR4 in migrating cortical interneurons. *Development*, *141*(9), 1857–1863. https://doi.org/10.1242/dev.104224.

Addison, C. L., Belperio, J. A., Burdick, M. D., & Strieter, R. M. (2004). Overexpression of the duffy antigen receptor for chemokines (DARC) by NSCLC tumor cells results in increased tumor necrosis. *BMC Cancer*, *4*, 28. https://doi.org/10.1186/1471-2407-4-28.

Adlere, I., Caspar, B., Arimont, M., Dekkers, S., Visser, K., Stuijt, J., et al. (2019). Modulators of CXCR4 and CXCR7/ACKR3 function. *Molecular Pharmacology*. https://doi.org/10.1124/mol.119.117663.

Ameti, R., Melgrati, S., Radice, E., Cameroni, E., Hub, E., Thelen, S., et al. (2018). Characterization of a chimeric chemokine as a specific ligand for ACKR3. *Journal of Leukocyte Biology*, *104*(2), 391–400. https://doi.org/10.1002/JLB.2MA1217-509R.

Bachelerie, F., Ben-Baruch, A., Burkhardt, A. M., Combadiere, C., Farber, J. M., Graham, G. J., et al. (2014). International Union of Basic and Clinical Pharmacology. [corrected]. LXXXIX. Update on the extended family of chemokine receptors and introducing a new nomenclature for atypical chemokine receptors. *Pharmacological Reviews*, *66*(1), 1–79. https://doi.org/10.1124/pr.113.007724.

Bachelerie, F., Graham, G. J., Locati, M., Mantovani, A., Murphy, P. M., Nibbs, R., et al. (2014). New nomenclature for atypical chemokine receptors. *Nature Immunology*, *15*(3), 207–208. https://doi.org/10.1038/ni.2812.
Bachelerie, F., Graham, G. J., Locati, M., Mantovani, A., Murphy, P. M., Nibbs, R., et al. (2015). An atypical addition to the chemokine receptor nomenclature: IUPHAR Review 15. *British Journal of Pharmacology*, *172*(16), 3945–3949. https://doi.org/10.1111/bph.13182.
Balabanian, K., Lagane, B., Infantino, S., Chow, K. Y., Harriague, J., Moepps, B., et al. (2005). The chemokine SDF-1/CXCL12 binds to and signals through the orphan receptor RDC1 in T lymphocytes. *The Journal of Biological Chemistry*, *280*(42), 35760–35766. https://doi.org/10.1074/jbc.M508234200.
Bandyopadhyay, S., Zhan, R., Chaudhuri, A., Watabe, M., Pai, S. K., Hirota, S., et al. (2006). Interaction of KAI1 on tumor cells with DARC on vascular endothelium leads to metastasis suppression. *Nature Medicine*, *12*(8), 933–938. https://doi.org/10.1038/nm1444.
Becker, J. H., Gao, Y., Soucheray, M., Pulido, I., Kikuchi, E., Rodriguez, M. L., et al. (2019). CXCR7 reactivates ERK signaling to promote resistance to EGFR kinase inhibitors in NSCLC. *Cancer Research*, *79*(17), 4439–4452. https://doi.org/10.1158/0008-5472.CAN-19-0024.
Behnam Azad, B., Lisok, A., Chatterjee, S., Poirier, J. T., Pullambhatla, M., Luker, G. D., et al. (2016). Targeted imaging of the atypical chemokine receptor 3 (ACKR3/CXCR7) in human cancer xenografts. *Journal of Nuclear Medicine*, *57*(6), 981–988. https://doi.org/10.2967/jnumed.115.167932.
Benhadjeba, S., Edjekouane, L., Sauve, K., Carmona, E., & Tremblay, A. (2018). Feedback control of the CXCR7/CXCL11 chemokine axis by estrogen receptor alpha in ovarian cancer. *Molecular Oncology*, *12*(10), 1689–1705. https://doi.org/10.1002/1878-0261.12362.
Berahovich, R. D., Zabel, B. A., Lewen, S., Walters, M. J., Ebsworth, K., Wang, Y., et al. (2014). Endothelial expression of CXCR7 and the regulation of systemic CXCL12 levels. *Immunology*, *141*(1), 111–122. https://doi.org/10.1111/imm.12176.
Berahovich, R. D., Zabel, B. A., Penfold, M. E., Lewen, S., Wang, Y., Miao, Z., et al. (2010). CXCR7 protein is not expressed on human or mouse leukocytes. *Journal of Immunology*, *185*(9), 5130–5139. https://doi.org/10.4049/jimmunol.1001660.
Bianco, A. M., Uno, M., Oba-Shinjo, S. M., Clara, C. A., de Almeida Galatro, T. F., Rosemberg, S., et al. (2015). CXCR7 and CXCR4 expressions in infiltrative astrocytomas and their interactions with HIF1alpha expression and IDH1 mutation. *Pathology Oncology Research*, *21*(2), 229–240. https://doi.org/10.1007/s12253-014-9813-7.
Birner, P., Tchorbanov, A., Natchev, S., Tuettenberg, J., & Guentchev, M. (2015). The chemokine receptor CXCR7 influences prognosis in human glioma in an IDH1-dependent manner. *Journal of Clinical Pathology*, *68*(10), 830–834. https://doi.org/10.1136/jclinpath-2015-202886.
Boehm, M., Beaumont, K., Jones, R., Kalgutkar, A. S., Zhang, L., Atkinson, K., et al. (2017). Discovery of potent and orally bioavailable macrocyclic peptide-peptoid hybrid CXCR7 modulators. *Journal of Medicinal Chemistry*, *60*(23), 9653–9663. https://doi.org/10.1021/acs.jmedchem.7b01028.
Boldajipour, B., Mahabaleshwar, H., Kardash, E., Reichman-Fried, M., Blaser, H., Minina, S., et al. (2008). Control of chemokine-guided cell migration by ligand sequestration. *Cell*, *132*(3), 463–473. https://doi.org/10.1016/j.cell.2007.12.034.
Bonecchi, R., Locati, M., Galliera, E., Vulcano, M., Sironi, M., Fra, A. M., et al. (2004). Differential recognition and scavenging of native and truncated macrophage-derived chemokine (macrophage-derived chemokine/CC chemokine ligand 22) by the D6 decoy receptor. *Journal of Immunology*, *172*(8), 4972–4976. https://doi.org/10.4049/jimmunol.172.8.4972.

Bonini, J. A., Martin, S. K., Dralyuk, F., Roe, M. W., Philipson, L. H., & Steiner, D. F. (1997). Cloning, expression, and chromosomal mapping of a novel human CC-chemokine receptor (CCR10) that displays high-affinity binding for MCP-1 and MCP-3. *DNA and Cell Biology*, *16*(10), 1249–1256. https://doi.org/10.1089/dna.1997.16.1249.

Borroni, E. M., Cancellieri, C., Vacchini, A., Benureau, Y., Lagane, B., Bachelerie, F., et al. (2013). Beta-arrestin-dependent activation of the cofilin pathway is required for the scavenging activity of the atypical chemokine receptor D6. *Science Signaling*, 6(273), S31–S33. https://doi.org/10.1126/scisignal.2003627. ra30 31–11.

Bunting, M. D., Comerford, I., Seach, N., Hammett, M. V., Asquith, D. L., Korner, H., et al. (2013). CCX-CKR deficiency alters thymic stroma impairing thymocyte development and promoting autoimmunity. *Blood*, *121*(1), 118–128. https://doi.org/10.1182/blood-2012-06-434886.

Burger, M., Hartmann, T., Krome, M., Rawluk, J., Tamamura, H., Fujii, N., et al. (2005). Small peptide inhibitors of the CXCR4 chemokine receptor (CD184) antagonize the activation, migration, and antiapoptotic responses of CXCL12 in chronic lymphocytic leukemia B cells. *Blood*, *106*(5), 1824–1830. https://doi.org/10.1182/blood-2004-12-4918.

Burns, J. M., Summers, B. C., Wang, Y., Melikian, A., Berahovich, R., Miao, Z., et al. (2006). A novel chemokine receptor for SDF-1 and I-TAC involved in cell survival, cell adhesion, and tumor development. *The Journal of Experimental Medicine*, *203*(9), 2201–2213. https://doi.org/10.1084/jem.20052144.

Calatozzolo, C., Canazza, A., Pollo, B., Di Pierro, E., Ciusani, E., Maderna, E., et al. (2011). Expression of the new CXCL12 receptor, CXCR7, in gliomas. *Cancer Biology & Therapy*, *11*(2), 242–253. https://doi.org/10.4161/cbt.11.2.13951.

Canals, M., Scholten, D. J., de Munnik, S., Han, M. K., Smit, M. J., & Leurs, R. (2012). Ubiquitination of CXCR7 controls receptor trafficking. *PLoS One*, 7(3). e34192https://doi.org/10.1371/journal.pone.0034192.

Catusse, J., Leick, M., Groch, M., Clark, D. J., Buchner, M. V., Zirlik, K., et al. (2010). Role of the atypical chemoattractant receptor CRAM in regulating CCL19 induced CCR7 responses in B-cell chronic lymphocytic leukemia. *Molecular Cancer*, *9*, 297. https://doi.org/10.1186/1476-4598-9-297.

Chakera, A., Seeber, R. M., John, A. E., Eidne, K. A., & Greaves, D. R. (2008). The duffy antigen/receptor for chemokines exists in an oligomeric form in living cells and functionally antagonizes CCR5 signaling through hetero-oligomerization. *Molecular Pharmacology*, *73*(5), 1362–1370. https://doi.org/10.1124/mol.107.040915.

Chaudhuri, A., Nielsen, S., Elkjaer, M. L., Zbrzezna, V., Fang, F., & Pogo, A. O. (1997). Detection of Duffy antigen in the plasma membranes and caveolae of vascular endothelial and epithelial cells of nonerythroid organs. *Blood*, *89*(2), 701–712.

Chaudhuri, A., Zbrzezna, V., Polyakova, J., Pogo, A. O., Hesselgesser, J., & Horuk, R. (1994). Expression of the Duffy antigen in K562 cells. Evidence that it is the human erythrocyte chemokine receptor. *The Journal of Biological Chemistry*, *269*(11), 7835–7838.

Chen, J., Yao, Y., Gong, C., Yu, F., Su, S., Chen, J., et al. (2011). CCL18 from tumor-associated macrophages promotes breast cancer metastasis via PITPNM3. *Cancer Cell*, *19*(4), 541–555. https://doi.org/10.1016/j.ccr.2011.02.006.

Chevigne, A., Fievez, V., Schmit, J. C., & Deroo, S. (2011). Engineering and screening the N-terminus of chemokines for drug discovery. *Biochemical Pharmacology*, *82*(10), 1438–1456. https://doi.org/10.1016/j.bcp.2011.07.091.

Coggins, N. L., Trakimas, D., Chang, S. L., Ehrlich, A., Ray, P., Luker, K. E., et al. (2014). CXCR7 controls competition for recruitment of beta-arrestin 2 in cells expressing both CXCR4 and CXCR7. *PLoS One*, *9*(6). e98328https://doi.org/10.1371/journal.pone.0098328.

Comerford, I., Milasta, S., Morrow, V., Milligan, G., & Nibbs, R. (2006). The chemokine receptor CCX-CKR mediates effective scavenging of CCL19 in vitro. *European Journal of Immunology*, *36*(7), 1904–1916. https://doi.org/10.1002/eji.200535716.

Comerford, I., Nibbs, R. J., Litchfield, W., Bunting, M., Harata-Lee, Y., Haylock-Jacobs, S., et al. (2010). The atypical chemokine receptor CCX-CKR scavenges homeostatic chemokines in circulation and tissues and suppresses Th17 responses. *Blood*, *116*(20), 4130–4140. https://doi.org/10.1182/blood-2010-01-264390.

Cutbush, M., Mollison, P. L., & Parkin, D. M. (1950). A new human blood group. *Nature*, *165*(4188), 188–189. https://doi.org/10.1038/165188b0.

D'Alterio, C., Consales, C., Polimeno, M., Franco, R., Cindolo, L., Portella, L., et al. (2010). Concomitant CXCR4 and CXCR7 expression predicts poor prognosis in renal cancer. *Current Cancer Drug Targets*, *10*(7), 772–781.

Darbonne, W. C., Rice, G. C., Mohler, M. A., Apple, T., Hebert, C. A., Valente, A. J., et al. (1991). Red blood cells are a sink for interleukin 8, a leukocyte chemotaxin. *The Journal of Clinical Investigation*, *88*(4), 1362–1369. https://doi.org/10.1172/JCI115442.

Datar, I., Sanmamed, M. F., Wang, J., Henick, B. S., Choi, J., Badri, T., et al. (2019). Expression analysis and significance of PD-1, LAG-3, and TIM-3 in human non-small cell lung cancer using spatially resolved and multiparametric single-cell analysis. *Clinical Cancer Research*, *25*(15), 4663–4673. https://doi.org/10.1158/1078-0432.CCR-18-4142.

Dawson, T. C., Lentsch, A. B., Wang, Z., Cowhig, J. E., Rot, A., Maeda, N., et al. (2000). Exaggerated response to endotoxin in mice lacking the Duffy antigen/receptor for chemokines (DARC). *Blood*, *96*(5), 1681–1684.

Dona, E., Barry, J. D., Valentin, G., Quirin, C., Khmelinskii, A., Kunze, A., et al. (2013). Directional tissue migration through a self-generated chemokine gradient. *Nature*, *503*(7475), 285–289. https://doi.org/10.1038/nature12635.

Donzella, G. A., Schols, D., Lin, S. W., Este, J. A., Nagashima, K. A., Maddon, P. J., et al. (1998). AMD3100, a small molecule inhibitor of HIV-1 entry via the CXCR4 co-receptor. *Nature Medicine*, *4*(1), 72–77. https://doi.org/10.1038/nm0198-072.

Dorf, M. E., Berman, M. A., Tanabe, S., Heesen, M., & Luo, Y. (2000). Astrocytes express functional chemokine receptors. *Journal of Neuroimmunology*, *111*(1–2), 109–121. https://doi.org/10.1016/s0165-5728(00)00371-4.

Du, J., Luan, J., Liu, H., Daniel, T. O., Peiper, S., Chen, T. S., et al. (2002). Potential role for Duffy antigen chemokine-binding protein in angiogenesis and maintenance of homeostasis in response to stress. *Journal of Leukocyte Biology*, *71*(1), 141–153.

Elson, J. K., Beebe-Dimmer, J. L., Morgenstern, H., Chilkuri, M., Blanchard, J., & Lentsch, A. B. (2011). The Duffy antigen/receptor for chemokines (DARC) and prostate-cancer risk among Jamaican men. *Journal of Immigrant and Minority Health*, *13*(1), 36–41. https://doi.org/10.1007/s10903-010-9330-z.

Feng, L. Y., Ou, Z. L., Wu, F. Y., Shen, Z. Z., & Shao, Z. M. (2009). Involvement of a novel chemokine decoy receptor CCX-CKR in breast cancer growth, metastasis and patient survival. *Clinical Cancer Research*, *15*(9), 2962–2970. https://doi.org/10.1158/1078-0432.CCR-08-2495.

Fievez, V., Szpakowska, M., Mosbah, A., Arumugam, K., Mathu, J., Counson, M., et al. (2018). Development of Mimokines, chemokine N terminus-based CXCR4 inhibitors optimized by phage display and rational design. *Journal of Leukocyte Biology*, *104*(2), 343–357. https://doi.org/10.1002/JLB.3MA0118-007.

Fra, A. M., Locati, M., Otero, K., Sironi, M., Signorelli, P., Massardi, M. L., et al. (2003). Cutting edge: Scavenging of inflammatory CC chemokines by the promiscuous putatively silent chemokine receptor D6. *Journal of Immunology*, *170*(5), 2279–2282. https://doi.org/10.4049/jimmunol.170.5.2279.

Freitas, C., Desnoyer, A., Meuris, F., Bachelerie, F., Balabanian, K., & Machelon, V. (2014). The relevance of the chemokine receptor ACKR3/CXCR7 on CXCL12-mediated

effects in cancers with a focus on virus-related cancers. *Cytokine & Growth Factor Reviews*, *25*(3), 307–316. https://doi.org/10.1016/j.cytogfr.2014.04.006.
Fukuma, N., Akimitsu, N., Hamamoto, H., Kusuhara, H., Sugiyama, Y., & Sekimizu, K. (2003). A role of the Duffy antigen for the maintenance of plasma chemokine concentrations. *Biochemical and Biophysical Research Communications*, *303*(1), 137–139. https://doi.org/10.1016/s0006-291x(03)00293-6.
Galliera, E., Jala, V. R., Trent, J. O., Bonecchi, R., Signorelli, P., Lefkowitz, R. J., et al. (2004). beta-Arrestin-dependent constitutive internalization of the human chemokine decoy receptor D6. *The Journal of Biological Chemistry*, *279*(24), 25590–25597. https://doi.org/10.1074/jbc.M400363200.
Gardner, L., Patterson, A. M., Ashton, B. A., Stone, M. A., & Middleton, J. (2004). The human Duffy antigen binds selected inflammatory but not homeostatic chemokines. *Biochemical and Biophysical Research Communications*, *321*(2), 306–312. https://doi.org/10.1016/j.bbrc.2004.06.146.
Goguet-Surmenian, E., Richard-Fiardo, P., Guillemot, E., Benchetrit, M., Gomez-Brouchet, A., Buzzo, P., et al. (2013). CXCR7-mediated progression of osteosarcoma in the lungs. *British Journal of Cancer*, *109*(6), 1579–1585. https://doi.org/10.1038/bjc.2013.482.
Goltsev, Y., Samusik, N., Kennedy-Darling, J., Bhate, S., Hale, M., Vazquez, G., et al. (2018). deep profiling of mouse splenic architecture with CODEX multiplexed imaging. *Cell*, *174*(4), 968–981. e915. https://doi.org/10.1016/j.cell.2018.07.010.
Gosling, J., Dairaghi, D. J., Wang, Y., Hanley, M., Talbot, D., Miao, Z., et al. (2000). Cutting edge: Identification of a novel chemokine receptor that binds dendritic cell- and T cell-active chemokines including ELC, SLC, and TECK. *Journal of Immunology*, *164*(6), 2851–2856. https://doi.org/10.4049/jimmunol.164.6.2851.
Graham, G. J., Locati, M., Mantovani, A., Rot, A., & Thelen, M. (2012). The biochemistry and biology of the atypical chemokine receptors. *Immunology Letters*, *145*(1–2), 30–38. https://doi.org/10.1016/j.imlet.2012.04.004.
Gravel, S., Malouf, C., Boulais, P. E., Berchiche, Y. A., Oishi, S., Fujii, N., et al. (2010). The peptidomimetic CXCR4 antagonist TC14012 recruits beta-arrestin to CXCR7: Roles of receptor domains. *The Journal of Biological Chemistry*, *285*(49), 37939–37943. https://doi.org/10.1074/jbc.C110.147470.
Guillemot, E., Karimdjee-Soilihi, B., Pradelli, E., Benchetrit, M., Goguet-Surmenian, E., Millet, M. A., et al. (2012). CXCR7 receptors facilitate the progression of colon carcinoma within lung not within liver. *British Journal of Cancer*, *107*(12), 1944–1949. https://doi.org/10.1038/bjc.2012.503.
Gustavsson, M., Wang, L., van Gils, N., Stephens, B. S., Zhang, P., Schall, T. J., et al. (2017). Structural basis of ligand interaction with atypical chemokine receptor 3. *Nature Communications*, *8*. 14135. https://doi.org/10.1038/ncomms14135.
Hansell, C. A. H., Fraser, A. R., Hayes, A. J., Pingen, M., Burt, C. L., Lee, K. M., et al. (2018). The atypical chemokine receptor Ackr2 constrains NK cell migratory activity and promotes metastasis. *Journal of Immunology*, *201*(8), 2510–2519. https://doi.org/10.4049/jimmunol.1800131.
Hansell, C. A., Schiering, C., Kinstrie, R., Ford, L., Bordon, Y., McInnes, I. B., et al. (2011). Universal expression and dual function of the atypical chemokine receptor D6 on innate-like B cells in mice. *Blood*, *117*(20), 5413–5424. https://doi.org/10.1182/blood-2010-11-317115.
Hao, M., Weng, X., Wang, Y., Sun, X., Yan, T., Li, Y., et al. (2018). Targeting CXCR7 improves the efficacy of breast cancer patients with tamoxifen therapy. *Biochemical Pharmacology*, *147*, 128–140. https://doi.org/10.1016/j.bcp.2017.11.013.
Harata-Lee, Y., Turvey, M. E., Brazzatti, J. A., Gregor, C. E., Brown, M. P., Smyth, M. J., et al. (2014). The atypical chemokine receptor CCX-CKR regulates metastasis of

mammary carcinoma via an effect on EMT. *Immunology and Cell Biology*, *92*(10), 815–824. https://doi.org/10.1038/icb.2014.58.

Hartmann, T. N., Grabovsky, V., Pasvolsky, R., Shulman, Z., Buss, E. C., Spiegel, A., et al. (2008). A crosstalk between intracellular CXCR7 and CXCR4 involved in rapid CXCL12-triggered integrin activation but not in chemokine-triggered motility of human T lymphocytes and CD34+ cells. *Journal of Leukocyte Biology*, *84*(4), 1130–1140. https://doi.org/10.1189/jlb.0208088.

Hassan, S., Ferrario, C., Saragovi, U., Quenneville, L., Gaboury, L., Baccarelli, A., et al. (2009). The influence of tumor-host interactions in the stromal cell-derived factor-1/ CXCR4 ligand/receptor axis in determining metastatic risk in breast cancer. *The American Journal of Pathology*, *175*(1), 66–73. https://doi.org/10.2353/ajpath.2009.080948.

Heinzel, K., Benz, C., & Bleul, C. C. (2007). A silent chemokine receptor regulates steady-state leukocyte homing in vivo. *Proceedings of the National Academy of Sciences of the United States of America*, *104*(20), 8421–8426. https://doi.org/10.1073/pnas.0608274104.

Hernandez, L., Magalhaes, M. A., Coniglio, S. J., Condeelis, J. S., & Segall, J. E. (2011). Opposing roles of CXCR4 and CXCR7 in breast cancer metastasis. *Breast Cancer Research*, *13*(6). R128https://doi.org/10.1186/bcr3074.

Horton, L. W., Yu, Y., Zaja-Milatovic, S., Strieter, R. M., & Richmond, A. (2007). Opposing roles of murine duffy antigen receptor for chemokine and murine CXC chemokine receptor-2 receptors in murine melanoma tumor growth. *Cancer Research*, *67*(20), 9791–9799. https://doi.org/10.1158/0008-5472.CAN-07-0246.

Horuk, R., Chitnis, C. E., Darbonne, W. C., Colby, T. J., Rybicki, A., Hadley, T. J., et al. (1993). A receptor for the malarial parasite Plasmodium vivax: The erythrocyte chemokine receptor. *Science*, *261*(5125), 1182–1184. https://doi.org/10.1126/science.7689250.

Hou, T., Liang, D., Xu, L., Huang, X., Huang, Y., & Zhang, Y. (2013). Atypical chemokine receptors predict lymph node metastasis and prognosis in patients with cervical squamous cell cancer. *Gynecologic Oncology*, *130*(1), 181–187. https://doi.org/10.1016/j.ygyno.2013.04.015.

Humpert, M. L., Tzouros, M., Thelen, S., Bignon, A., Levoye, A., Arenzana-Seisdedos, F., et al. (2012). Complementary methods provide evidence for the expression of CXCR7 on human B cells. *Proteomics*, *12*(12), 1938–1948. https://doi.org/10.1002/pmic.201100581.

Ierano, C., Santagata, S., Napolitano, M., Guardia, F., Grimaldi, A., Antignani, E., et al. (2014). CXCR4 and CXCR7 transduce through mTOR in human renal cancer cells. *Cell Death & Disease*, *5*, e1310. https://doi.org/10.1038/cddis.2014.269.

Ikeda, Y., Kumagai, H., Skach, A., Sato, M., & Yanagisawa, M. (2013). Modulation of circadian glucocorticoid oscillation via adrenal opioid-CXCR7 signaling alters emotional behavior. *Cell*, *155*(6), 1323–1336. https://doi.org/10.1016/j.cell.2013.10.052.

Jamieson, T., Cook, D. N., Nibbs, R. J., Rot, A., Nixon, C., McLean, P., et al. (2005). The chemokine receptor D6 limits the inflammatory response in vivo. *Nature Immunology*, *6*(4), 403–411. https://doi.org/10.1038/ni1182.

Juarez, J., Bradstock, K. F., Gottlieb, D. J., & Bendall, L. J. (2003). Effects of inhibitors of the chemokine receptor CXCR4 on acute lymphoblastic leukemia cells in vitro. *Leukemia*, *17*(7), 1294–1300. https://doi.org/10.1038/sj.leu.2402998.

Kalatskaya, I., Berchiche, Y. A., Gravel, S., Limberg, B. J., Rosenbaum, J. S., & Heveker, N. (2009). AMD3100 is a CXCR7 ligand with allosteric agonist properties. *Molecular Pharmacology*, *75*(5), 1240–1247. https://doi.org/10.1124/mol.108.053389.

Keeley, E. C., Mehrad, B., & Strieter, R. M. (2010). CXC chemokines in cancer angiogenesis and metastases. *Advances in Cancer Research*, *106*, 91–111. https://doi.org/10.1016/S0065-230X(10)06003-3.

Khanna, P., Chung, C. Y., Neves, R. I., Robertson, G. P., & Dong, C. (2014). CD82/KAI expression prevents IL-8-mediated endothelial gap formation in late-stage melanomas. *Oncogene*, *33*(22), 2898–2908. https://doi.org/10.1038/onc.2013.249.

Kin, N. W., Crawford, D. M., Liu, J., Behrens, T. W., & Kearney, J. F. (2008). DNA microarray gene expression profile of marginal zone versus follicular B cells and idiotype positive marginal zone B cells before and after immunization with Streptococcus pneumoniae. *Journal of Immunology*, *180*(10), 6663–6674. https://doi.org/10.4049/jimmunol.180.10.6663.

Klein, K. R., Karpinich, N. O., Espenschied, S. T., Willcockson, H. H., Dunworth, W. P., Hoopes, S. L., et al. (2014). Decoy receptor CXCR7 modulates adrenomedullin-mediated cardiac and lymphatic vascular development. *Developmental Cell*, *30*(5), 528–540. https://doi.org/10.1016/j.devcel.2014.07.012.

Kleist, A. B., Getschman, A. E., Ziarek, J. J., Nevins, A. M., Gauthier, P. A., Chevigne, A., et al. (2016). New paradigms in chemokine receptor signal transduction: Moving beyond the two-site model. *Biochemical Pharmacology*, *114*, 53–68. https://doi.org/10.1016/j.bcp.2016.04.007.

Lacalle, R. A., Blanco, R., Carmona-Rodriguez, L., Martin-Leal, A., Mira, E., & Manes, S. (2017). Chemokine receptor signaling and the hallmarks of cancer. *International Review of Cell and Molecular Biology*, *331*, 181–244. https://doi.org/10.1016/bs.ircmb.2016.09.011.

Latini, F. R., Bastos, A. U., Arnoni, C. P., Muniz, J. G., Person, R. M., Baleotti, W., Jr., et al. (2013). DARC (Duffy) and BCAM (Lutheran) reduced expression in thyroid cancer. *Blood Cells, Molecules & Diseases*, *50*(3), 161–165. https://doi.org/10.1016/j.bcmd.2012.10.009.

Lazennec, G., & Richmond, A. (2010). Chemokines and chemokine receptors: New insights into cancer-related inflammation. *Trends in Molecular Medicine*, *16*(3), 133–144. https://doi.org/10.1016/j.molmed.2010.01.003.

Lee, J. S., Frevert, C. W., Wurfel, M. M., Peiper, S. C., Wong, V. A., Ballman, K. K., et al. (2003). Duffy antigen facilitates movement of chemokine across the endothelium in vitro and promotes neutrophil transmigration in vitro and in vivo. *Journal of Immunology*, *170*(10), 5244–5251. https://doi.org/10.4049/jimmunol.170.10.5244.

Lee, K. M., McKimmie, C. S., Gilchrist, D. S., Pallas, K. J., Nibbs, R. J., Garside, P., et al. (2011). D6 facilitates cellular migration and fluid flow to lymph nodes by suppressing lymphatic congestion. *Blood*, *118*(23), 6220–6229. https://doi.org/10.1182/blood-2011-03-344044.

Lee, K. M., Wilson, G. J., Pingen, M., Fukuoka, A., Hansell, C. A. H., Bartolini, R., et al. (2019). Placental chemokine compartmentalisation: A novel mammalian molecular control mechanism. *PLoS Biology*, *17*(5). e3000287https://doi.org/10.1371/journal.pbio.3000287.

Lee, J. S., Wurfel, M. M., Matute-Bello, G., Frevert, C. W., Rosengart, M. R., Ranganathan, M., et al. (2006). The Duffy antigen modifies systemic and local tissue chemokine responses following lipopolysaccharide stimulation. *Journal of Immunology*, *177*(11), 8086–8094. https://doi.org/10.4049/jimmunol.177.11.8086.

Leick, M., Catusse, J., Follo, M., Nibbs, R. J., Hartmann, T. N., Veelken, H., et al. (2010). CCL19 is a specific ligand of the constitutively recycling atypical human chemokine receptor CRAM-B. *Immunology*, *129*(4), 536–546. https://doi.org/10.1111/j.1365-2567.2009.03209.x.

Levoye, A., Balabanian, K., Baleux, F., Bachelerie, F., & Lagane, B. (2009). CXCR7 heterodimerizes with CXCR4 and regulates CXCL12-mediated G protein signaling. *Blood*, *113*(24), 6085–6093. https://doi.org/10.1182/blood-2008-12-196618.

Li, S., Fong, K. W., Gritsina, G., Zhang, A., Zhao, J. C., Kim, J., et al. (2019). Activation of MAPK signaling by CXCR7 leads to enzalutamide resistance in prostate cancer. *Cancer Research*, *79*(10), 2580–2592. https://doi.org/10.1158/0008-5472.CAN-18-2812.

Li, J. T., Jia, L. T., Liu, N. N., Zhu, X. S., Liu, Q. Q., Wang, X. L., et al. (2015). MiRNA-101 inhibits breast cancer growth and metastasis by targeting CX chemokine receptor 7. *Oncotarget*, *6*(31), 30818–30830. https://doi.org/10.18632/oncotarget.5067.

Li, D. D., Yang, C., Shao, Z. M., & Yu, K. D. (2018). Effect of functional genetic variants in chemokine decoy receptors on the recurrence risk of breast cancer. *Cancer Medicine*, 7(11), 5497–5504. https://doi.org/10.1002/cam4.1823.

Lin, Z., Li, W., Zhang, H., Wu, W., Peng, Y., Zeng, Y., et al. (2016). CCL18/PITPNM3 enhances migration, invasion, and EMT through the NF-kappaB signaling pathway in hepatocellular carcinoma. *Tumour Biology*, *37*(3), 3461–3468. https://doi.org/10.1007/s13277-015-4172-x.

Liu, Y., Zheng, H., Li, Q., Li, S., Lai, H., Song, E., et al. (2019). Discovery of CCL18 antagonist blocking breast cancer metastasis. *Clinical & Experimental Metastasis*, *36*(3), 243–255. https://doi.org/10.1007/s10585-019-09965-2.

Luker, K. E., Lewin, S. A., Mihalko, L. A., Schmidt, B. T., Winkler, J. S., Coggins, N. L., et al. (2012). Scavenging of CXCL12 by CXCR7 promotes tumor growth and metastasis of CXCR4-positive breast cancer cells. *Oncogene*, *31*(45), 4750–4758. https://doi.org/10.1038/onc.2011.633.

Luker, K. E., Steele, J. M., Mihalko, L. A., Ray, P., & Luker, G. D. (2010). Constitutive and chemokine-dependent internalization and recycling of CXCR7 in breast cancer cells to degrade chemokine ligands. *Oncogene*, *29*(32), 4599–4610. https://doi.org/10.1038/onc.2010.212.

Luo, Y., Azad, A. K., Karanika, S., Basourakos, S. P., Zuo, X., Wang, J., et al. (2018). Enzalutamide and CXCR7 inhibitor combination treatment suppresses cell growth and angiogenic signaling in castration-resistant prostate cancer models. *International Journal of Cancer*, *142*(10), 2163–2174. https://doi.org/10.1002/ijc.31237.

Maeda, S., Kuboki, S., Nojima, H., Shimizu, H., Yoshitomi, H., Furukawa, K., et al. (2017). Duffy antigen receptor for chemokines (DARC) expressing in cancer cells inhibits tumor progression by suppressing CXCR2 signaling in human pancreatic ductal adenocarcinoma. *Cytokine*, *95*, 12–21. https://doi.org/10.1016/j.cyto.2017.02.007.

Maishi, N., Ohga, N., Hida, Y., Akiyama, K., Kitayama, K., Osawa, T., et al. (2012). CXCR7: A novel tumor endothelial marker in renal cell carcinoma. *Pathology International*, *62*(5), 309–317. https://doi.org/10.1111/j.1440-1827.2012.02792.x.

Martinez de la Torre, Y., Buracchi, C., Borroni, E. M., Dupor, J., Bonecchi, R., Nebuloni, M., et al. (2007). Protection against inflammation- and autoantibody-caused fetal loss by the chemokine decoy receptor D6. *Proceedings of the National Academy of Sciences of the United States of America*, *104*(7), 2319–2324. https://doi.org/10.1073/pnas.0607514104.

Martinez de la Torre, Y., Locati, M., Buracchi, C., Dupor, J., Cook, D. N., Bonecchi, R., et al. (2005). Increased inflammation in mice deficient for the chemokine decoy receptor D6. *European Journal of Immunology*, *35*(5), 1342–1346. https://doi.org/10.1002/eji.200526114.

Massara, M., Bonavita, O., Mantovani, A., Locati, M., & Bonecchi, R. (2016). Atypical chemokine receptors in cancer: Friends or foes? *Journal of Leukocyte Biology*, *99*(6), 927–933. https://doi.org/10.1189/jlb.3MR0915-431RR.

Massara, M., Bonavita, O., Savino, B., Caronni, N., Mollica Poeta, V., Sironi, M., et al. (2018). ACKR2 in hematopoietic precursors as a checkpoint of neutrophil release and anti-metastatic activity. *Nature Communications*, *9*(1). 676https://doi.org/10.1038/s41467-018-03080-8.

Maussang, D., Mujic-Delic, A., Descamps, F. J., Stortelers, C., Vanlandschoot, P., Stigter-van Walsum, M., et al. (2013). Llama-derived single variable domains (nanobodies) directed against chemokine receptor CXCR7 reduce head and neck cancer cell growth in vivo. *The Journal of Biological Chemistry*, *288*(41), 29562–29572. https://doi.org/10.1074/jbc.M113.498436.

Mayr, F. B., Spiel, A. O., Leitner, J. M., Firbas, C., Kliegel, T., Jilma-Stohlawetz, P., et al. (2008). Duffy antigen modifies the chemokine response in human endotoxemia. *Critical Care Medicine, 36*(1), 159–165. https://doi.org/10.1097/01.CCM.0000297875.55969.DB.

McKimmie, C. S., Fraser, A. R., Hansell, C., Gutierrez, L., Philipsen, S., Connell, L., et al. (2008). Hemopoietic cell expression of the chemokine decoy receptor D6 is dynamic and regulated by GATA1. *Journal of Immunology, 181*(11), 8171–8181. https://doi.org/10.4049/jimmunol.181.11.8170-a.

Miao, Z., Luker, K. E., Summers, B. C., Berahovich, R., Bhojani, M. S., Rehemtulla, A., et al. (2007). CXCR7 (RDC1) promotes breast and lung tumor growth in vivo and is expressed on tumor-associated vasculature. *Proceedings of the National Academy of Sciences of the United States of America, 104*(40), 15735–15740. https://doi.org/10.1073/pnas.0610444104.

Miller, L. H., Mason, S. J., Clyde, D. F., & McGinniss, M. H. (1976). The resistance factor to Plasmodium vivax in blacks. The Duffy-blood-group genotype, FyFy. *The New England Journal of Medicine, 295*(6), 302–304. https://doi.org/10.1056/NEJM197608052950602.

Nalls, M. A., Wilson, J. G., Patterson, N. J., Tandon, A., Zmuda, J. M., Huntsman, S., et al. (2008). Admixture mapping of white cell count: Genetic locus responsible for lower white blood cell count in the Health ABC and Jackson Heart studies. *American Journal of Human Genetics, 82*(1), 81–87. https://doi.org/10.1016/j.ajhg.2007.09.003.

Naumann, U., Cameroni, E., Pruenster, M., Mahabaleshwar, H., Raz, E., Zerwes, H. G., et al. (2010). CXCR7 functions as a scavenger for CXCL12 and CXCL11. *PLoS One, 5*(2). e9175https://doi.org/10.1371/journal.pone.0009175.

Neote, K., Mak, J. Y., Kolakowski, L. F. ,. J., & Schall, T. J. (1994). Functional and biochemical analysis of the cloned Duffy antigen: Identity with the red blood cell chemokine receptor. *Blood, 84*(1), 44–52.

Nibbs, R. J., & Graham, G. J. (2013). Immune regulation by atypical chemokine receptors. *Nature Reviews. Immunology, 13*(11), 815–829.

Nibbs, R. J., Kriehuber, E., Ponath, P. D., Parent, D., Qin, S., Campbell, J. D., et al. (2001). The beta-chemokine receptor D6 is expressed by lymphatic endothelium and a subset of vascular tumors. *The American Journal of Pathology, 158*(3), 867–877. https://doi.org/10.1016/s0002-9440(10)64035-7.

Nibbs, R. J., Wylie, S. M., Pragnell, I. B., & Graham, G. J. (1997). Cloning and characterization of a novel murine beta chemokine receptor, D6. Comparison to three other related macrophage inflammatory protein-1alpha receptors, CCR-1, CCR-3, and CCR-5. *The Journal of Biological Chemistry, 272*(19), 12495–12504. https://doi.org/10.1074/jbc.272.19.12495.

Nibbs, R. J., Wylie, S. M., Yang, J., Landau, N. R., & Graham, G. J. (1997). Cloning and characterization of a novel promiscuous human beta-chemokine receptor D6. *The Journal of Biological Chemistry, 272*(51), 32078–32083. https://doi.org/10.1074/jbc.272.51.32078.

Nomiyama, H., Osada, N., & Yoshie, O. (2011). A family tree of vertebrate chemokine receptors for a unified nomenclature. *Developmental and Comparative Immunology, 35*(7), 705–715. https://doi.org/10.1016/j.dci.2011.01.019.

Odemis, V., Lipfert, J., Kraft, R., Hajek, P., Abraham, G., Hattermann, K., et al. (2012). The presumed atypical chemokine receptor CXCR7 signals through G(i/o) proteins in primary rodent astrocytes and human glioma cells. *Glia, 60*(3), 372–381. https://doi.org/10.1002/glia.22271.

Orimo, A., Gupta, P. B., Sgroi, D. C., Arenzana-Seisdedos, F., Delaunay, T., Naeem, R., et al. (2005). Stromal fibroblasts present in invasive human breast carcinomas promote tumor growth and angiogenesis through elevated SDF-1/CXCL12 secretion. *Cell, 121*(3), 335–348. https://doi.org/10.1016/j.cell.2005.02.034.

Pacifico, F., Lepore, A., Mellone, S., Sanguigno, L., Federico, G., Greco, A., et al. (2017). The chemokine scavenging receptor D6/ACKR2 is a target of miR-146a in thyroid cancer. *Genes & Cancer, 8*(5–6), 577–588. https://doi.org/10.18632/genesandcancer.141.

Peiper, S. C., Wang, Z. X., Neote, K., Martin, A. W., Showell, H. J., Conklyn, M. J., et al. (1995). The Duffy antigen/receptor for chemokines (DARC) is expressed in endothelial cells of Duffy negative individuals who lack the erythrocyte receptor. *The Journal of Experimental Medicine, 181*(4), 1311–1317. https://doi.org/10.1084/jem.181.4.1311.

Pruenster, M., Mudde, L., Bombosi, P., Dimitrova, S., Zsak, M., Middleton, J., et al. (2009). The Duffy antigen receptor for chemokines transports chemokines and supports their promigratory activity. *Nature Immunology, 10*(1), 101–108. https://doi.org/10.1038/ni.1675.

Puddinu, V., Casella, S., Radice, E., Thelen, S., Dirnhofer, S., Bertoni, F., et al. (2017). ACKR3 expression on diffuse large B cell lymphoma is required for tumor spreading and tissue infiltration. *Oncotarget, 8*(49), 85068–85084. https://doi.org/10.18632/oncotarget.18844.

Qiao, Y., Zhang, C., Li, A., Wang, D., Luo, Z., Ping, Y., et al. (2018). IL6 derived from cancer-associated fibroblasts promotes chemoresistance via CXCR7 in esophageal squamous cell carcinoma. *Oncogene, 37*(7), 873–883. https://doi.org/10.1038/onc.2017.387.

Rajagopal, S., Kim, J., Ahn, S., Craig, S., Lam, C. M., Gerard, N. P., et al. (2010). Beta-arrestin- but not G protein-mediated signaling by the "decoy" receptor CXCR7. *Proceedings of the National Academy of Sciences of the United States of America, 107*(2), 628–632. https://doi.org/10.1073/pnas.0912852107.

Regard, J. B., Sato, I. T., & Coughlin, S. R. (2008). Anatomical profiling of G protein-coupled receptor expression. *Cell, 135*(3), 561–571. https://doi.org/10.1016/j.cell.2008.08.040.

Reich, D., Nalls, M. A., Kao, W. H., Akylbekova, E. L., Tandon, A., Patterson, N., et al. (2009). Reduced neutrophil count in people of African descent is due to a regulatory variant in the Duffy antigen receptor for chemokines gene. *PLoS Genetics, 5*(1). e1000360https://doi.org/10.1371/journal.pgen.1000360.

Rot, A., McKimmie, C., Burt, C. L., Pallas, K. J., Jamieson, T., Pruenster, M., et al. (2013). Cell-autonomous regulation of neutrophil migration by the D6 chemokine decoy receptor. *Journal of Immunology, 190*(12), 6450–6456. https://doi.org/10.4049/jimmunol.1201429.

Salazar, N., Carlson, J. C., Huang, K., Zheng, Y., Oderup, C., Gross, J., et al. (2018). A chimeric antibody against ACKR3/CXCR7 in combination with TMZ activates immune responses and extends survival in mouse GBM models. *Molecular Therapy, 26*(5), 1354–1365. https://doi.org/10.1016/j.ymthe.2018.02.030.

Salazar, N., Munoz, D., Kallifatidis, G., Singh, R. K., Jorda, M., & Lokeshwar, B. L. (2014). The chemokine receptor CXCR7 interacts with EGFR to promote breast cancer cell proliferation. *Molecular Cancer, 13*, 198. https://doi.org/10.1186/1476-4598-13-198.

Sanchez-Alcaniz, J. A., Haege, S., Mueller, W., Pla, R., Mackay, F., Schulz, S., et al. (2011). Cxcr7 controls neuronal migration by regulating chemokine responsiveness. *Neuron, 69*(1), 77–90. https://doi.org/10.1016/j.neuron.2010.12.006.

Sanchez-Martin, L., Sanchez-Mateos, P., & Cabanas, C. (2013). CXCR7 impact on CXCL12 biology and disease. *Trends in Molecular Medicine, 19*(1), 12–22. https://doi.org/10.1016/j.molmed.2012.10.004.

Savas, P., Virassamy, B., Ye, C., Salim, A., Mintoff, C. P., Caramia, F., et al. (2018). Single-cell profiling of breast cancer T cells reveals a tissue-resident memory subset associated with improved prognosis. *Nature Medicine, 24*(7), 986–993. https://doi.org/10.1038/s41591-018-0078-7.

Savino, B., Caronni, N., Anselmo, A., Pasqualini, F., Borroni, E. M., Basso, G., et al. (2014). ERK-dependent downregulation of the atypical chemokine receptor D6 drives tumor

aggressiveness in Kaposi sarcoma. *Cancer Immunology Research*, *2*(7), 679–689. https://doi.org/10.1158/2326-6066.CIR-13-0202.
Schall, T. J., & Proudfoot, A. E. (2011). Overcoming hurdles in developing successful drugs targeting chemokine receptors. *Nature Reviews. Immunology*, *11*(5), 355–363. https://doi.org/10.1038/nri2972.
Schnabel, R. B., Baumert, J., Barbalic, M., Dupuis, J., Ellinor, P. T., Durda, P., et al. (2010). Duffy antigen receptor for chemokines (Darc) polymorphism regulates circulating concentrations of monocyte chemoattractant protein-1 and other inflammatory mediators. *Blood*, *115*(26), 5289–5299. https://doi.org/10.1182/blood-2009-05-221382.
Schols, D., Este, J. A., Henson, G., & De Clercq, E. (1997). Bicyclams, a class of potent anti-HIV agents, are targeted at the HIV coreceptor fusin/CXCR-4. *Antiviral Research*, *35*(3), 147–156. https://doi.org/10.1016/s0166-3542(97)00025-9.
Schrevel, M., Karim, R., ter Haar, N. T., van der Burg, S. H., Trimbos, J. B., Fleuren, G. J., et al. (2012). CXCR7 expression is associated with disease-free and disease-specific survival in cervical cancer patients. *British Journal of Cancer*, *106*(9), 1520–1525. https://doi.org/10.1038/bjc.2012.110.
Shams, K., Kurowska-Stolarska, M., Schutte, F., Burden, A. D., McKimmie, C. S., & Graham, G. J. (2018). MicroRNA-146 and cell trauma down-regulate expression of the psoriasis-associated atypical chemokine receptor ACKR2. *The Journal of Biological Chemistry*, *293*(8), 3003–3012. https://doi.org/10.1074/jbc.M117.809780.
Shen, H., Schuster, R., Stringer, K. F., Waltz, S. E., & Lentsch, A. B. (2006). The Duffy antigen/receptor for chemokines (DARC) regulates prostate tumor growth. *The FASEB Journal*, *20*(1), 59–64. https://doi.org/10.1096/fj.05-4764com.
Shi, J. Y., Yang, L. X., Wang, Z. C., Wang, L. Y., Zhou, J., Wang, X. Y., et al. (2015). CC chemokine receptor-like 1 functions as a tumour suppressor by impairing CCR7-related chemotaxis in hepatocellular carcinoma. *The Journal of Pathology*, *235*(4), 546–558. https://doi.org/10.1002/path.4450.
Shin, W. J., Zabel, B. A., & Pachynski, R. K. (2018). Mechanisms and functions of chemerin in cancer: Potential roles in therapeutic intervention. *Frontiers in Immunology*, *9*, 2772. https://doi.org/10.3389/fimmu.2018.02772.
Sierro, F., Biben, C., Martinez-Munoz, L., Mellado, M., Ransohoff, R. M., Li, M., et al. (2007). Disrupted cardiac development but normal hematopoiesis in mice deficient in the second CXCL12/SDF-1 receptor, CXCR7. *Proceedings of the National Academy of Sciences of the United States of America*, *104*(37), 14759–14764. https://doi.org/10.1073/pnas.0702229104.
Sjoberg, E., Meyrath, M., Milde, L., Herrera, M., Lovrot, J., Hagerstrand, D., et al. (2019). A novel ACKR2-dependent role of fibroblast-derived CXCL14 in epithelial-to-mesenchymal transition and metastasis of breast cancer. *Clinical Cancer Research*, *25*(12), 3702–3717. https://doi.org/10.1158/1078-0432.CCR-18-1294.
Stacer, A. C., Fenner, J., Cavnar, S. P., Xiao, A., Zhao, S., Chang, S. L., et al. (2016). Endothelial CXCR7 regulates breast cancer metastasis. *Oncogene*, *35*(13), 1716–1724. https://doi.org/10.1038/onc.2015.236.
Su, A. I., Cooke, M. P., Ching, K. A., Hakak, Y., Walker, J. R., Wiltshire, T., et al. (2002). Large-scale analysis of the human and mouse transcriptomes. *Proceedings of the National Academy of Sciences of the United States of America*, *99*(7), 4465–4470. https://doi.org/10.1073/pnas.012025199.
Sun, X., Cheng, G., Hao, M., Zheng, J., Zhou, X., Zhang, J., et al. (2010). CXCL12/CXCR4/CXCR7 chemokine axis and cancer progression. *Cancer Metastasis Reviews*, *29*(4), 709–722. https://doi.org/10.1007/s10555-010-9256-x.
Sun, G., Wang, Y., Zhu, Y., Huang, C., & Ji, Q. (2011). Duffy antigen receptor for chemokines in laryngeal squamous cell carcinoma as a negative regulator. *Acta Oto-Laryngologica*, *131*(2), 197–203. https://doi.org/10.3109/00016489.2010.516012.

Szabo, M. C., Soo, K. S., Zlotnik, A., & Schall, T. J. (1995). Chemokine class differences in binding to the Duffy antigen-erythrocyte chemokine receptor. *The Journal of Biological Chemistry*, *270*(43), 25348–25351. https://doi.org/10.1074/jbc.270.43.25348.

Szpakowska, M., & Chevigne, A. (2016). vCCL2/vMIP-II, the viral master KEYmokine. *Journal of Leukocyte Biology*, *99*(6), 893–900. https://doi.org/10.1189/jlb.2MR0815-383R.

Szpakowska, M., Dupuis, N., Baragli, A., Counson, M., Hanson, J., Piette, J., et al. (2016). Human herpesvirus 8-encoded chemokine vCCL2/vMIP-II is an agonist of the atypical chemokine receptor ACKR3/CXCR7. *Biochemical Pharmacology*, *114*, 14–21. https://doi.org/10.1016/j.bcp.2016.05.012.

Szpakowska, M., Meyrath, M., Reynders, N., Counson, M., Hanson, J., Steyaert, J., et al. (2018). Mutational analysis of the extracellular disulphide bridges of the atypical chemokine receptor ACKR3/CXCR7 uncovers multiple binding and activation modes for its chemokine and endogenous non-chemokine agonists. *Biochemical Pharmacology*, *153*, 299–309. https://doi.org/10.1016/j.bcp.2018.03.007.

Szpakowska, M., Nevins, A. M., Meyrath, M., Rhainds, D., D'Huys, T., Guite-Vinet, F., et al. (2018). Different contributions of chemokine N-terminal features attest to a different ligand binding mode and a bias towards activation of ACKR3/CXCR7 compared with CXCR4 and CXCR3. *British Journal of Pharmacology*, *175*(9), 1419–1438. https://doi.org/10.1111/bph.14132.

Thelen, M., & Thelen, S. (2008). CXCR7, CXCR4 and CXCL12: An eccentric trio? *Journal of Neuroimmunology*, *198*(1–2), 9–13. https://doi.org/10.1016/j.jneuroim.2008.04.020.

Thiriot, A., Perdomo, C., Cheng, G., Novitzky-Basso, I., McArdle, S., Kishimoto, J. K., et al. (2017). Differential DARC/ACKR1 expression distinguishes venular from non-venular endothelial cells in murine tissues. *BMC Biology*, *15*(1). 45https://doi.org/10.1186/s12915-017-0381-7.

Townson, J. R., & Nibbs, R. J. (2002). Characterization of mouse CCX-CKR, a receptor for the lymphocyte-attracting chemokines TECK/mCCL25, SLC/mCCL21 and MIP-3beta/mCCL19: Comparison to human CCX-CKR. *European Journal of Immunology*, *32*(5), 1230–1241. https://doi.org/10.1002/1521-4141(200205)32:5<1230::AID-IMMU1230>3.0.CO;2-L.

Ulvmar, M. H., Werth, K., Braun, A., Kelay, P., Hub, E., Eller, K., et al. (2014). The atypical chemokine receptor CCRL1 shapes functional CCL21 gradients in lymph nodes. *Nature Immunology*, *15*(7), 623–630. https://doi.org/10.1038/ni.2889.

Walenkamp, A. M. E., Lapa, C., Herrmann, K., & Wester, H. J. (2017). CXCR4 ligands: The next big hit? *Journal of Nuclear Medicine*, *58*(Suppl 2), 77S–82S. https://doi.org/10.2967/jnumed.116.186874.

Wang, H., Beaty, N., Chen, S., Qi, C. F., Masiuk, M., Shin, D. M., et al. (2012). The CXCR7 chemokine receptor promotes B-cell retention in the splenic marginal zone and serves as a sink for CXCL12. *Blood*, *119*(2), 465–468. https://doi.org/10.1182/blood-2011-03-343608.

Wang, Y., Li, G., Stanco, A., Long, J. E., Crawford, D., Potter, G. B., et al. (2011). CXCR4 and CXCR7 have distinct functions in regulating interneuron migration. *Neuron*, *69*(1), 61–76. https://doi.org/10.1016/j.neuron.2010.12.005.

Wang, J., Ou, Z. L., Hou, Y. F., Luo, J. M., Shen, Z. Z., Ding, J., et al. (2006). Enhanced expression of Duffy antigen receptor for chemokines by breast cancer cells attenuates growth and metastasis potential. *Oncogene*, *25*(54), 7201–7211. https://doi.org/10.1038/sj.onc.1209703.

Wang, J., Shiozawa, Y., Wang, J., Wang, Y., Jung, Y., Pienta, K. J., et al. (2008). The role of CXCR7/RDC1 as a chemokine receptor for CXCL12/SDF-1 in prostate cancer. *The Journal of Biological Chemistry*, *283*(7), 4283–4294. https://doi.org/10.1074/jbc.M707465200.

Wani, N., Nasser, M. W., Ahirwar, D. K., Zhao, H., Miao, Z., Shilo, K., et al. (2014). C-X-C motif chemokine 12/C-X-C chemokine receptor type 7 signaling regulates breast cancer growth and metastasis by modulating the tumor microenvironment. *Breast Cancer Research*, *16*(3), R54. https://doi.org/10.1186/bcr3665.

Weber, M., Blair, E., Simpson, C. V., O'Hara, M., Blackburn, P. E., Rot, A., et al. (2004). The chemokine receptor D6 constitutively traffics to and from the cell surface to internalize and degrade chemokines. *Molecular Biology of the Cell*, *15*(5), 2492–2508. https://doi.org/10.1091/mbc.e03-09-0634.

Werner, T. A., Forster, C. M., Dizdar, L., Verde, P. E., Raba, K., Schott, M., et al. (2017). CXCR4/CXCR7/CXCL12 axis promotes an invasive phenotype in medullary thyroid carcinoma. *British Journal of Cancer*, *117*(12), 1837–1845. https://doi.org/10.1038/bjc.2017.364.

Wijtmans, M., Maussang, D., Sirci, F., Scholten, D. J., Canals, M., Mujic-Delic, A., et al. (2012). Synthesis, modeling and functional activity of substituted styrene-amides as small-molecule CXCR7 agonists. *European Journal of Medicinal Chemistry*, *51*, 184–192. https://doi.org/10.1016/j.ejmech.2012.02.041.

Wu, F. Y., Ou, Z. L., Feng, L. Y., Luo, J. M., Wang, L. P., Shen, Z. Z., et al. (2008). Chemokine decoy receptor d6 plays a negative role in human breast cancer. *Molecular Cancer Research*, *6*(8), 1276–1288. https://doi.org/10.1158/1541-7786.MCR-07-2108.

Wu, W., Qian, L., Chen, X., & Ding, B. (2015). Prognostic significance of CXCL12, CXCR4, and CXCR7 in patients with breast cancer. *International Journal of Clinical and Experimental Pathology*, *8*(10), 13217–13224.

Wu, Y. C., Tang, S. J., Sun, G. H., & Sun, K. H. (2016). CXCR7 mediates TGFbeta1-promoted EMT and tumor-initiating features in lung cancer. *Oncogene*, *35*(16), 2123–2132. https://doi.org/10.1038/onc.2015.274.

Yamada, K., Maishi, N., Akiyama, K., Towfik Alam, M., Ohga, N., Kawamoto, T., et al. (2015). CXCL12-CXCR7 axis is important for tumor endothelial cell angiogenic property. *International Journal of Cancer*, *137*(12), 2825–2836. https://doi.org/10.1002/ijc.29655.

Yang, D., Dai, T., Xue, L., Liu, X., Wu, B., Geng, J., et al. (2015). Expression of chemokine receptor CXCR7 in colorectal carcinoma and its prognostic significance. *International Journal of Clinical and Experimental Pathology*, *8*(10), 13051–13058.

Yang, C., Yu, K. D., Xu, W. H., Chen, A. X., Fan, L., Ou, Z. L., et al. (2013). Effect of genetic variants in two chemokine decoy receptor genes, DARC and CCBP2, on metastatic potential of breast cancer. *PLoS One*, *8*(11). e78901https://doi.org/10.1371/journal.pone.0078901.

Yu, S., Crawford, D., Tsuchihashi, T., Behrens, T. W., & Srivastava, D. (2011). The chemokine receptor CXCR7 functions to regulate cardiac valve remodeling. *Developmental Dynamics*, *240*(2), 384–393. https://doi.org/10.1002/dvdy.22549.

Yu, P. F., Huang, Y., Xu, C. L., Lin, L. Y., Han, Y. Y., Sun, W. H., et al. (2017). Downregulation of CXCL12 in mesenchymal stromal cells by TGFbeta promotes breast cancer metastasis. *Oncogene*, *36*(6), 840–849. https://doi.org/10.1038/onc.2016.252.

Yu, K. D., Wang, X., Yang, C., Zeng, X. H., & Shao, Z. M. (2015). Host genotype and tumor phenotype of chemokine decoy receptors integrally affect breast cancer relapse. *Oncotarget*, *6*(28), 26519–26527. https://doi.org/10.18632/oncotarget.4470.

Zabel, B. A., Wang, Y., Lewen, S., Berahovich, R. D., Penfold, M. E., Zhang, P., et al. (2009). Elucidation of CXCR7-mediated signaling events and inhibition of CXCR4-mediated tumor cell transendothelial migration by CXCR7 ligands. *Journal of Immunology*, *183*(5), 3204–3211. https://doi.org/10.4049/jimmunol.0900269.

Zeng, X. H., Ou, Z. L., Yu, K. D., Feng, L. Y., Yin, W. J., Li, J., et al. (2011). Coexpression of atypical chemokine binders (ACBs) in breast cancer predicts better outcomes. *Breast Cancer Research and Treatment*, *125*(3), 715–727. https://doi.org/10.1007/s10549-010-0875-2.

Zheng, N., Liu, W., Chen, J., Li, B., Liu, J., Wang, J., et al. (2019). CXCR7 is not obligatory for CXCL12-CXCR4-induced epithelial-mesenchymal transition in human ovarian cancer. *Molecular Carcinogenesis*, *58*(1), 144–155. https://doi.org/10.1002/mc.22916.

Zhou, S., Liu, M., Hu, Y., An, W., Liang, X., Yu, W., et al. (2015). Expression of Duffy antigen receptor for chemokines (DARC) is down-regulated in colorectal cancer. *Journal of Receptor and Signal Transduction Research*, *35*(5), 462–467. https://doi.org/10.3109/10799893.2015.1009113.

Zhu, Q., Jiang, L., & Wang, X. (2017). The expression of Duffy antigen receptor for chemokines by epithelial ovarian cancer decreases growth potential. *Oncology Letters*, *13*(6), 4302–4306. https://doi.org/10.3892/ol.2017.5954.

Zhu, Z., Sun, Z., Wang, Z., Guo, P., Zheng, X., & Xu, H. (2013). Prognostic impact of atypical chemokine receptor expression in patients with gastric cancer. *The Journal of Surgical Research*, *183*(1), 177–183. https://doi.org/10.1016/j.jss.2013.01.023.

Zhu, Y., Tang, W., Liu, Y., Wang, G., Liang, Z., & Cui, L. (2014). CCX-CKR expression in colorectal cancer and patient survival. *The International Journal of Biological Markers*, *29*(1), e40–e48. https://doi.org/10.5301/jbm.5000057.

Zlotnik, A., & Yoshie, O. (2000). Chemokines: A new classification system and their role in immunity. *Immunity*, *12*(2), 121–127. https://doi.org/10.1016/s1074-7613(00)80165-x.

Zlotnik, A., & Yoshie, O. (2012). The chemokine superfamily revisited. *Immunity*, *36*(5), 705–716. https://doi.org/10.1016/j.immuni.2012.05.008.

CHAPTER FIVE

Structure and function of β-arrestins, their emerging role in breast cancer, and potential opportunities for therapeutic manipulation

Arun K. Shukla*, Hemlata Dwivedi-Agnihotri
Department of Biological Sciences and Bioengineering, Indian Institute of Technology, Kanpur, India
*Corresponding author: e-mail address: arshukla@iitk.ac.in

Contents

Abstract

β-Arrestins (βarrs) are multifunctional intracellular proteins with an ability to directly interact with a large number of cellular partners including the G protein-coupled receptors (GPCRs). βarrs contribute to multiple aspects of GPCR signaling, trafficking and downregulation. Considering the central involvement of GPCR signaling in the onset and progression of diverse types of cancers, βarrs have also emerged as key players in the context of investigating cancer phenotypes, and as potential therapeutic targets. In this chapter, we first provide a brief account of structure and function of βarrs and then highlight recent discoveries unfolding novel functional attributes of βarrs in breast cancer. We also underscore the recent paradigms of modulating βarr functions in cellular context and potential therapeutic opportunities going forward.

Advances in Cancer Research, Volume 145
ISSN 0065-230X
https://doi.org/10.1016/bs.acr.2020.01.001

1. Introduction

β-Arrestins (βarrs) are cytoplasmic proteins with ubiquitous expression throughout the body, and due to their critical role in regulating GPCRs, they have emerged as one of the important nodes in cellular signaling pathways (Lefkowitz & Shenoy, 2005). In the classical paradigm of GPCR signaling and regulation, βarrs are used as one of the primary mechanisms to terminate heterotrimeric G-protein coupling to agonist-bound GPCRs through steric hindrance based mechanism (DeWire, Ahn, Lefkowitz, & Shenoy, 2007). Their roles, however, have broadened significantly over the last two decades with uncovering of an ever-increasing number of interactions that they are involved in, and fundamental cellular processes that they directly or indirectly regulate downstream of GPCRs (Gurevich & Gurevich, 2019; Kang, Tian, & Benovic, 2014; Shenoy & Lefkowitz, 2011) (Fig. 1).

For example, βarrs interact with various components of clathrin coated endocytosis machinery to regulate agonist-induced receptor trafficking (Kang et al., 2014). Not only it serves as a mechanism for downregulating receptor density at the cell surface and thereby downstream signaling response but it also drives receptor compartmentalization to influence functional outcomes (Calebiro, Godbole, Lyga, & Lohse, 2015; Lobingier & von Zastrow, 2019; Vilardaga, Jean-Alphonse, & Gardella, 2014).

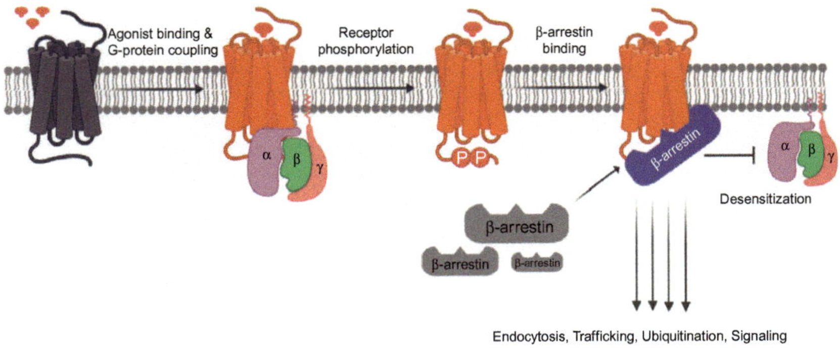

Fig. 1 A schematic representation showing the multifaceted role of β-arrestins in GPCR signaling and regulation. Agonist-stimulation leads to a conformational change in GPCRs followed by the interaction and activation of heterotrimeric G-proteins. Subsequently, GPCRs are phosphorylated by GPCR kinases (GRKs) that facilitate the binding of β-arrestins. GPCR-β-arrestin interaction terminates further G-protein coupling via steric hindrance mechanism on one hand while on the other, it initiates receptor internalization and β-arrestin mediated signaling.

Similarly, βarrs can also interact with E3 ubiquitin ligases to mediate receptor ubiquitination and degradation (Shenoy & Lefkowitz, 2011), and a diverse set of kinases and phosphatases to contribute to downstream signaling (DeWire et al., 2007; Peterson & Luttrell, 2017). More recently, formation of GPCR-G-protein-βarr complexes have also been described as a potential mechanism for endosomal signaling by GPCRs (Thomsen et al., 2016).

A number of studies across different GPCRs have established a contribution of βarrs in various cellular processes including cell cycle regulation, cellular proliferation and migration which are directly linked with the onset and development of different types of cancer (Bagnato & Rosano, 2019). As a result, investigating the role of βarrs in multiple aspects of carcinogenesis and cancer metastasis has come to the forefront, especially in the context of GPCR signaling. In addition to βarrs, related proteins referred to as Arrestin Domain Containing Proteins (ARRDCs) (Aubry & Klein, 2013) have also been implicated in different aspects of cancer phenotypes in *ex vivo* and *in vivo* model systems.

2. Structure and function of β-arrestins

As mentioned above, βarrs mediate a broad spectrum of functional outcomes in the context of GPCR signaling and regulation. In terms of their relevance in breast cancer, and cancer in general, the ability of βarrs to influence receptor trafficking and contribute to downstream signaling are most critical. The subfamily of arrestins includes four different members (named as arrestin 1–4) of which, two (i.e., arrestin 1 and 4) are primarily restricted to visual system, and they are typically referred to as visual-arrestins. The role of visual arrestins is limited primarily to the visual receptor, rhodopsin. The other two subtypes of arrestins, more commonly known as β-arrestin 1 (arrestin-2) and β-arrestin 2 (arrestin-3) are ubiquitously distributed and they typically interact with, and modulate the functions of, the majority of GPCRs.

βarrs have a two-domain structure (i.e., the N- and the C-domain), which primarily consists of anti-parallel β-strands linked with small loop regions (Gurevich & Gurevich, 2019). The carboxyl-terminus of βarrs is folded back onto the N-domain and contributes in maintaining βarrs in the basal state. The basal conformation of βarrs is stabilized by two different intramolecular interactions referred to as the "polar core" and the "three-element interaction" which are disrupted upon their interaction with activated and phosphorylated GPCRs (Gurevich & Gurevich, 2019).

Multiple structures of βarrs are now described in literature including an active conformation of βarr1 in complex with a phosphorylated peptide corresponding to the carboxyl-terminus of the vasopressin V2 receptor (Shukla et al., 2013). In addition, a low-resolution architecture of a chimeric β2 adrenergic receptor with βarr1 (Shukla, Westfield, et al., 2014) and a cryo-EM structure of the neurotensin receptor-βarr1 fusion protein (Yin et al., 2019) have also been described. These studies have started to provide direct structural insights into GPCR-βarr interaction and activation mechanisms although high-resolution structural details on GPCR-βarr complexes are still awaited.

The two isoforms of βarrs are structurally similar but often display functional divergence in the context of GPCR signaling and regulation (Srivastava, Gupta, Gupta, & Shukla, 2015). A recent study has demonstrated distinct conformations of receptor-bound βarr1 and 2 as a mechanism for their functional divergence (Ghosh et al., 2019). In addition, there are also emerging indications of preferential recruitment of one isoform over the other for different receptor systems. Furthermore, the conformational diversity sampled by βarrs upon their interaction with differentially phosphorylated receptors adds an additional level of functional fine-tuning which is conceptualized in terms of a bar code hypothesis (Nobles et al., 2011; Reiter & Lefkowitz, 2006; Shukla et al., 2008) and a flute model (Yang et al., 2017). Finally, distinct binding modalities in receptor-βarr complexes have also been linked to different functional outcomes, which further broaden their functional capabilities (Cahill et al., 2017; Kumari et al., 2016, 2017; Sente et al., 2018).

3. β-Arrestin interactome: Clues into their multi-functionality

A key feature of βarrs is their ability to bind a large number of cellular proteins and facilitate the formation of multi-protein signalosomes in cellular context (DeWire et al., 2007). The interaction network of βarrs are responsible for driving their broad functional capabilities to a large extent including their highly conserved role in GPCR internalization and signaling. In addition to focused studies on specific βarr interactions, a global proteomics analysis has identified a large number of interaction partners of βarrs upon stimulation of the angiotensin II type 1a receptor (Xiao et al., 2007). Interestingly, this study has identified a number of cellular partners of βarrs which are critical players in cell cycle progression, cellular proliferation and migration, further corroborating the potential role of βarrs in different types of cancer phenotypes (Xiao et al., 2007). This interactome of βarrs has paved the way for characterization of several novel interactions including those which are directly implicated in the context of carcinogenesis and metastasis relevant signaling pathways (Hara et al., 2011; Kovacs et al., 2008). An interesting study identified βarr interactome upon activation of PAR-2 in MCF-7 cells using mass spectrometry based proteomics approach (Parisis, Metodieva, & Metodiev, 2013). Similar to the βarr interactome mentioned above, this study also discovered a large number of βarr interaction partners, and subsequent bioinformatics analysis revealed that networks involved in interleukin signaling, actin cytoskeleton and PI3K/AKT pathways are significantly enriched (Parisis et al., 2013). This comprehensive study offers an extremely useful platform to probe the contribution of βarr interaction network in different aspects of breast cancer.

The latest addition to the list of βarr binding proteins is the interaction between βarr2 and MELK (Maternal embryonic leucine zipper kinase) (Perry et al., 2019). MELK expression is upregulated in a range of cancer cells including breast cancer cells. This study uncovers, using *in vitro* binding assay with purified proteins, that βarr2 directly interacts with the kinase domain of MELK, and moreover, also validates their interaction in cellular system (Perry et al., 2019). Most interestingly, co-expression of βarr2 and MELK decreases the number of cells in the S-phase of cell cycle suggesting a potential role of this interaction in cellular proliferation (Perry et al., 2019).

In addition to GPCRs, βarrs have also been found to impart regulatory effect on other membrane proteins including non-GPCR receptors, ion-channels and transporters (Chen et al., 2016; Kashihara, Nakada, Kojima, Takeshita, & Yamada, 2017; Lefkowitz, Rajagopal, & Whalen, 2006; Liu et al., 2017; Shenoy & Lefkowitz, 2011). For example, βarr1 plays a critical role in ubiquitination and downregulation of TRPV4 (Transient Receptor Potential cation channel subfamily V member 4) by bringing an E3 ubiquitin ligase to close proximity of AT1aR-TRPV4 heterodimer in the membrane (Shukla et al., 2010). Thus, it is plausible that βarrs play a significantly broader role in mediating the permeation of ions, and transport of substrates, than currently appreciated, which may have a significant bearing on various attributes of cellular outcomes relevant to cancer.

4. Emerging role of β-arrestins in cancer phenotype

There are a large number of reports suggesting a direct contribution of βarrs in diverse types of cancer phenotypes and they appear to be involved in nearly every stage including initiation, promotion and progression. While some of these involve GPCRs, others are mediated through non-GPCR mechanisms as well. A number of recent reviews have summarized an overview of various studies implicating βarrs in cancer (Bagnato & Rosano, 2019; Crudden et al., 2019; Sobolesky & Moussa, 2013; Tocci, Rosano, & Bagnato, 2019) and therefore, we will not describe those examples here. Instead, we have specifically focused the following subsection on the emerging contribution of βarrs in breast cancer with a particular emphasis on studies described in the last 5 years or so.

5. Role of β-arrestins in breast cancer

Several independent studies have reported a change in βarr expression pattern in breast cancer cells and tissues, and also identified critical roles of βarrs in signaling pathways and cellular outcomes implicated in breast cancer phenotype (Table 1). Interestingly, majority of cell line based studies are carried out in TNBC (Triple Negative Breast Cancer) cells, which are relatively more aggressive with poorer outcomes and higher recurrence. For example, βarrs are required for PAR-2 (Protease Activated Receptor subtype 2) mediated migration of MDA MB-231 breast cancer cell lines potentially via an ERK1/2 MAP kinase pathway (Ge et al., 2004). In MCF-7 and MDA MB-231 cells, βarr2 appears to play an inhibitory role in morphine-induced apoptosis via an Akt and caspase-8 pathway (Zhao et al., 2009). Similarly, βarrs have been implicated in the migration and invasion of HS578T and MDA-MB-231 cell lines in context of the lysophosphatidic acid (LPA) receptor through a member of the Ras GTPase family referred to as Ral (Li et al., 2009). Moreover, several studies have implicated the so-called non-canonical GPCRs, also referred to as atypical chemokine receptors (ACKRs), in breast cancer phenotype (Massara, Bonavita, Mantovani, Locati, & Bonecchi, 2016; Mollica Poeta, Massara, Capucetti, & Bonecchi, 2019; Wang, Chen, & Shen, 2018). This is particularly relevant considering that these receptors do not couple to heterotrimeric G-proteins but robustly recruit and activate βarrs (Bachelerie, Ben-Baruch, et al., 2014; Bachelerie, Graham, et al., 2014). We discuss some of the recent studies providing novel aspects of βarr function in breast cancer and emerging mechanistic insights in the following section (Table 1).

Key signaling networks that βarrs appear to regulate involve ERK1/2 MAP kinase activation (DeWire et al., 2007), EGFR transactivation (Barki-Harrington & Rockman, 2008) and cytoskeletal rearrangements (Xiao et al., 2010). Interestingly, this arm of βarr function appears to be

Table 1 A brief summary of the contribution of β-arrestins in breast cancer phenotype as described in recent studies.

Protein (β-arrestin 1/2)	Expression/functional contribution/ localization	References
βarr1 and βarr2	Contribute in PAR-2 mediated migration of MDA MB-231 cells via ERK1/2 MAP kinase activation	Ge, Shenoy, Lefkowitz, and DeFea (2004)
βarr2	Significantly reduces morphine-induced cell death in MCF-7 and MDA-MB231 cells via an Akt and caspase-8 pathways	Zhao et al. (2009)
βarr1 and βarr2	Impair LPA-induced migration and invasion of MDA-MB-231 cells via a Ral GTPase mechanism	Li et al. (2009)
βarr1	Comparatively higher expression in ERα+ invasive ductal carcinoma compared to invasive ERα− ductal carcinomas	Rezaul et al. (2010)
βarr1	Inhibits the migration of MDA-MB-468 and MDA-MB-231 cells	Lundgren et al. (2011)
βarr2	Contributes in Kisspeptin-induced EGFR transactivation, and invasion of MDA-MB-231 cells via MMP-9	Zajac et al. (2011)
βarr2	Overexpression inhibits proliferation and promotes apoptosis of MCF-7 and T-47D cells via $GABA_BR$/JNK pathway	Wu, Shan, Zheng, and Pei (2014)
βarr2	Potentially contributes in GPR161-mediated proliferation and migration of MDA-MB-361 cells via IQGAP1 dependent mechanism	Feigin, Xue, Hammell, and Muthuswamy (2014)
βarr2	βarr2 expression correlates with the levels of MDR1-gp in breast tissue samples and contributes to doxorubicin sensitivity in MDA-MB-231 and MCF-7 cells	Jing et al. (2015)
βarr2	Contributes to Kisspeptin-induced invadopodia formation in MDA-MB-231 via ERK1/2 MAP kinase	Goertzen, Dragan, Turley, Babwah, and Bhattacharya (2016)
βarr1	βarr1 is involved in cellular proliferation and migration in TNBC cells, potentially via AMP kinase activation, and their expression is downregulated in TNBC cells; miR-374a-5p targets and regulates βarr1 expression in TNBC cells	Son et al. (2019)
βarr1 and βarr2	Proteomics-based βarr interactome in MCF-7 cells upon PAR2 activation	Parisis et al. (2013)

Please note that the table is not exhaustive and highlights some of the key examples available in the literature.

one of the key mechanisms through which they play a role in breast cancer, and other cancer types as well. Similar to PAR-2 example mentioned above, βarr2 is involved in kisspeptin receptor dependent formation of invadopodia in breast cancer cells via an EGFR and ERK1/2 activation pathway (Goertzen et al., 2016). The kisspeptin 1 receptor (KISS1R), also known as GPR54, is a prototypical GPCR that is activated by a peptide hormone called kisspeptin, previously known as metastin, that was originally identified as a human metastasis suppressor gene (Kirby, Maguire, Colledge, & Davenport, 2010). Invadopodia are membrane protrusions rich in actin that are often formed by cancer cells and they are involved in invasion properties and metastasis potential of these cells (Yamaguchi, 2012). Goertzen et al. observed that silencing βarr2 in MDA-MB-231 and Hs578T cell lines significantly inhibits kisspeptin-induced invadopodia formation, suggesting a potential involvement of βarr2 in invasion and metastatic potential of these cells.

miRNAs are single stranded, non-protein coding RNAs of 19–25 nucleotides, and they target mRNAs through a RISC (RNA-induced silencing complex) dependent mechanism (O'Brien, Hayder, Zayed, & Peng, 2018). Emerging data positions them as one of the major contributing factors in diverse types of cancer phenotypes and make them an attractive target for therapeutics (To et al., 2020). A recent study has identified miRNA based control of βarr1 expression in TNBC samples (Son et al., 2019). In a microarray screen, the authors observed a significant upregulation of miR-374a-5p in samples derived from TNBC patients (Son et al., 2019). In line with this observation, targeted inhibition of miR-374a-5p in MDM-MB-231, MDM-MB-468 and MDA-MB-157 cells impairs their growth and migration (Son et al., 2019). Interestingly, suppression of miR-374a-5p enhances βarr1 expression at mRNA and protein levels in these cell lines as well as in xenograft mouse model (Son et al., 2019). Exogenous overexpression of βarr1 in these cells decreases their survival, proliferation and migration potentially via an AMPK (AMP-activated protein kinase) dependent mechanism. Furthermore, the analysis of TCGA (The Cancer Genome Atlas) and METABRIC (Molecular Taxonomy of Breast Cancer International Consortium) datasets reveal that βarr1 expression is downregulated in TNBC patient samples, further underscoring its role as a potential tumor suppressor in breast cancer (Son et al., 2019).

Wu et al. have reported that in MCF-7 and T-47D breast cancer cell lines, the levels of βarr2 is lower compared to corresponding control cell lines, and βarr2 overexpression inhibits proliferation of these cells and triggers apoptosis (Wu et al., 2014). The contribution of βarr2 in this context involves GABA-B receptor and activation of JNK kinase (Wu et al., 2014). Another study has observed overexpression of an orphan GPCR, Gpr161, in MDA-MB-361 and BT-474 cells (Feigin et al., 2014). Knocking down Gpr161 decreases the proliferation of mammary epithelial cells while overexpression promotes proliferation and migration (Feigin et al., 2014). Interestingly, βarr2 appears to scaffold a complex of Gpr161 with IQGAP1 (IQ motif containing GTPase activating protein 1), which appears to be an important driving mechanism underlying Gpr161 mediated effects (Feigin et al., 2014).

One of the key problems associated with therapeutic management of breast cancer is the emergence of "Multi Drug Resistance" (MDR) (Martin, Smith, & Tomlinson, 2014). This is attributed, in part, to overexpression of a membrane protein MDR1/p-gp, which is an efflux transporter (also known as the ATP-binding cassette subfamily B member 1; ABCB1 or cluster of differentiation 243; CD243) (Endicott & Ling, 1989). Jing et al. analyzed approximately one hundred samples from breast cancer tissues in terms of βarr2 and MDR1 expression pattern, and observed a fairly significant correlation (Jing et al., 2015). Interestingly, βarr2 and MDR1 levels appear to be upregulated in MCF-7/ADM, a multidrug resistant cell line, compared to regular MCF-7 and MDA-MB-231 cells (Jing et al., 2015). Corroborating these findings, not only exogenous overexpression of βarr2 enhances MDR1 expression in MCF-7 and MDA-MB-231 cells but also siRNA mediated silencing of βarr2 results in a dramatic reduction in MDR1 expression in MCF-7/ADM cell line (Jing et al., 2015). Most importantly, silencing βarr2 increases doxorubicin sensitivity in these cell lines while its overexpression has an opposite effect (Jing et al., 2015). Taken together, these observations imply an important contribution of βarr2 in breast cancer MDR, and offer a potential avenue for therapeutic consideration. Considering the role of βarrs in fine-tuning surface levels of non-GPCR membrane proteins by mediating their internalization and/or ubiquitination, it is tempting to speculate that they may have a similar role in the context of MDR1 although it remains to be experimentally tested.

In addition to βarrs, Arrestin Domain Containing Proteins (ARRDCs) which are classified under a broad umbrella of arrestin-family of proteins (Aubry & Klein, 2013), have also emerged as key players in various aspects of breast cancer phenotypes (Arakaki, Pan, Lin, & Trejo, 2018; Soung et al., 2017; Soung, Pruitt, & Chung, 2014; Zheng et al., 2017).

ARRDC3 in particular appears to be involved at multiple levels, for example, in the reversal of chemo-resistance, invasive potential of extracellular vesicles derived from MCF-7 and MDM-MB-231 cells, PAR-1 induced MDA-MB-231 invasion, and proliferation and migration of MDM-MB-231 cells (Draheim et al., 2010; Soung, Chung, Yan, Ju, & Chung, 2019; Soung, Ford, Yan, & Chung, 2018; Soung et al., 2017, 2014). A comprehensive discussion of ARRDCs and their role in breast cancer requires a chapter of its own, and thus, it is not covered here in detail.

6. Potential strategies to modulate β-arrestin functions

Unlike GPCRs, βarrs are cytosolic proteins and therefore, strategies to target them directly, for example, using small molecules, pose significant challenges (Chaturvedi et al., 2018; Shukla, 2014). Moreover, universal inhibitors of βarrs such as those for enzymes may not be optimal considering their multi-functionality especially with respect to terminating G-protein signaling, receptor trafficking and signaling. Directly targeting βarrs is beginning to emerge as a niche area, and we discuss some of the recent reports in this section (Fig. 2).

As mentioned earlier, one of the key functions of βarrs is to mediate agonist-induced receptor endocytosis via scaffolding various players of

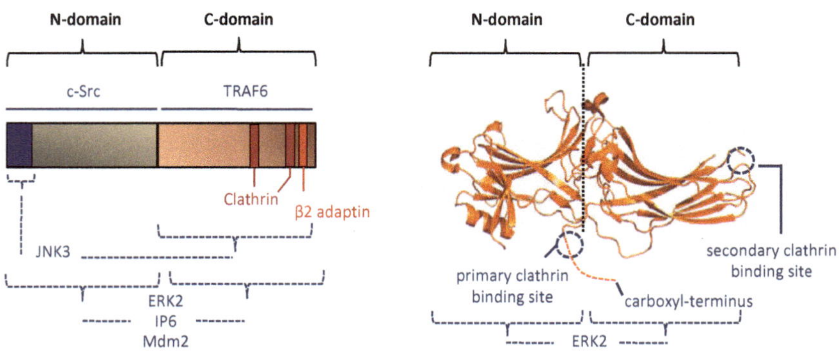

Fig. 2 Structural representation of β-arrestin indicating some of the interaction interfaces. A potential mechanism underlying the multi-functionality of β-arrestins is their ability to interact with a large number of cellular partners. As some of the interaction interfaces are quite distinct from each other, it offers a framework to modulate selected β-arrestin functions by targeting the protein-protein interaction. *This image is adopted from and modified based on a previous publication from our laboratory Ghosh, E., Srivastava, A., Baidya, M., Kumari, P., Dwivedi, H., Nidhi, K., et al. (2017). A synthetic intrabody-based selective and generic inhibitor of GPCR endocytosis. Nature Nanotechnology, 12, 1190–1198.*

the clathrin coated machinery including clathrin itself and β2-adaptin (Goodman et al., 1996, 1998; Laporte et al., 1999). Such a mechanism is important for controlling GPCR signaling from the perspective of receptor desensitization in the classical paradigm, as well as in terms of the emerging concept of GPCR signaling from endosomal compartments (Tsvetanova, Irannejad, & von Zastrow, 2015; Vilardaga et al., 2014). A recent study has used virtual screening approach to identify a small molecule compound named barbadin that is capable of inhibiting βarr-β2-adaptin interaction in cellular context (Beautrait et al., 2017). Although barbadin binds to β2-adpatin, it robustly inhibits agonist-induced endocytosis of a broad set of GPCRs by interfering with βarr-β2-adaptin interaction without significantly affecting GPCR-βarr interaction (Beautrait et al., 2017). Barbadin also inhibits sustained cAMP response arising from endosomes and agonist-induced ERK1/2 MAP kinase phosphorylation for selected GPCRs. Although not tested directly in the context of cancer phenotype, the discovery and characterization of barbadin provides a proof-of-principle for the conceptual framework of modulating βarr functions by targeting their interaction interfaces (Fig. 2).

A previous study has used a βarr2-targeting RNA aptamer in the context of chronic myelogenous leukemia (CML) (Jonathan et al., 2014) where βarr2 is critically involved in the onset and maintenance of chronic and blast crisis phases (Fereshteha et al., 2012). This study used an *in vitro* platform referred to as "Systematic Evolution of Ligands by EXponential enrichment (SELEX)" (Gold, 1995; Que-Gewirth & Sullenger, 2007) for selecting RNA aptamers against purified βarr2 and identified a number of different aptamers with reasonable binding affinities (Jonathan et al., 2014). Selected aptamers targeting βarr2 inhibited its interaction with ERK2 at submicromolar concentrations in pull-down assays and therefore, offered a possibility of disrupting βarr signaling in cellular context (Jonathan et al., 2014). Subsequently, βarr2 aptamer was linked to a previously described nucleolin targeting DNA aptamer (Kotula et al., 2012) which allowed its delivery to the K562 cell line which are immortalized myelogenous leukemia cells (Jonathan et al., 2014). Interestingly, this approach resulted in the disruption of multiple signaling pathways in these cells, potentially by interfering with βarr interactions, and also led to a significant reduction in cell growth (Jonathan et al., 2014). Although not tested in the context of breast cancer, it may be an attractive avenue to explore in future, especially considering that βarr-ERK axis appears to be a key player in proliferation and migration of breast cancer cells.

More recently, a set of synthetic antibody fragments (FABs) against βarrs were isolated from a phage display library and characterized in terms of their selectivity against βarr isoforms and ability to modulate different interactions of βarrs (Ghosh et al., 2017). One of these FABs targeting βarr2, referred to as FAB5, selectively inhibited the interaction of βarr2 with clathrin terminal domain *in vitro* (Ghosh et al., 2017). Upon expression as an intrabody, this particular FAB, now referred to as intrabody5, significantly inhibited agonist-induced endocytosis of a diverse set of GPCRs (Ghosh et al., 2017). This example of intrabody-based modulation of βarr interaction and its ensuing functional outcome provides a potential framework for targeting βarr signaling in the context of cancer therapeutics. Moreover, new platforms for generating different types of synthetic binders including nanobodies and monobodies have now been designed (Koide, Wojcik, Gilbreth, Hoey, & Koide, 2012; McMahon et al., 2018) which offer additional possibilities of targeting βarrs from a therapeutic standpoint.

An interesting avenue that has not been explored yet in the context of breast cancer, and cancer phenotype in general, is modulating βarr functions through biased agonism at GPCRs (Roy, Getschman, Volkman, & Dwinell, 2017). The concept of biased agonism broadly refers to preferential signaling through a specific transducer pathway over others, and it has emerged as a novel framework for improving GPCR targeted therapeutics (Costa-Neto, Parreiras, & Bouvier, 2016; Shukla, Singh, & Ghosh, 2014; Shukla, Xiao, & Lefkowitz, 2011). It is conceivable that a G-protein biased agonist may have therapeutic advantages over balanced agonists where βarr signaling contributes positively toward the onset and progression of cancer phenotype. Considering multiple studies demonstrating the role of βarrs in cellular proliferation, migration and invasion of breast cancer cells, this aspect of biased agonism may be highly relevant from therapeutic point of view. A recent review article has discussed this particular aspect in the context of chemokine receptors (Roy et al., 2017).

7. Conclusion and future perspective

Here, we have provided a brief overview of structure and function of βarrs and summarized the recent discoveries on the contribution of βarrs in breast cancer phenotype with specific case examples. We have also discussed several proof-of-principle studies that are now described to target βarr function in cellular context, including an aptamer-based strategy in a myeloid leukemia cell line. A key focus going forward should be on designing additional tools and reagents for directly targeting βarrs in cellular systems as well as in animal models. These future studies may provide promising avenues for therapeutic manipulation of βarr signaling in different cancer phenotypes in including breast cancer.

Acknowledgments

The research program in Dr. Shukla's laboratory is supported by an Intermediate Fellowship of the Wellcome Trust/DBT India Alliance Fellowship (grant number IA/I/14/1/501285); the Swarnajayanti Fellowship of Department of Science and Technology (DST/SJF/LSA-03/2017-18); the Science and Engineering Research Board (SERB) (EMR/2017/003804); Innovative Young Biotechnologist Award from the Department of Biotechnology (DBT) (BT/08/IYBA/2014-3); and the Indian Institute of Technology, Kanpur. A.K.S. is an Intermediate Fellow of Wellcome Trust/DBT India Alliance, EMBO Young Investigator, and Joy Gill Chair Professor. H.D.-A. is supported the National Post-Doctoral Fellowship of SERB (PDF/2016/002930) and DBT-BioCaRE grant (BT/PR31791/BIC/101/1228/2019). We apologize to the authors whose relevant work may have skipped our attention or could not be cited because of space limitation.

References

Arakaki, A. K. S., Pan, W. A., Lin, H., & Trejo, J. (2018). The alpha-arrestin ARRDC3 suppresses breast carcinoma invasion by regulating G protein-coupled receptor lysosomal sorting and signaling. *The Journal of Biological Chemistry*, *293*, 3350–3362.

Aubry, L., & Klein, G. (2013). True arrestins and arrestin-fold proteins: A structure-based appraisal. *Progress in Molecular Biology and Translational Science*, *118*, 21–56.

Bachelerie, F., Ben-Baruch, A., Burkhardt, A. M., Combadiere, C., Farber, J. M., Graham, G. J., et al. (2014). International Union of Basic and Clinical Pharmacology. [corrected]. LXXXIX. Update on the extended family of chemokine receptors and introducing a new nomenclature for atypical chemokine receptors. *Pharmacological Reviews*, *66*, 1–79.

Bachelerie, F., Graham, G. J., Locati, M., Mantovani, A., Murphy, P. M., Nibbs, R., et al. (2014). New nomenclature for atypical chemokine receptors. *Nature Immunology*, *15*, 207–208.

Bagnato, A., & Rosano, L. (2019). New routes in GPCR/beta-arrestin-driven signaling in cancer progression and metastasis. *Frontiers in Pharmacology*, *10*, 114.

Barki-Harrington, L., & Rockman, H. A. (2008). Beta-arrestins: Multifunctional cellular mediators. *Physiology (Bethesda, Md.)*, *23*, 17–22.

Beautrait, A., Paradis, J. S., Zimmerman, B., Giubilaro, J., Nikolajev, L., Armando, S., et al. (2017). A new inhibitor of the beta-arrestin/AP2 endocytic complex reveals interplay between GPCR internalization and signalling. *Nature Communications*, *8*, 15054.

Cahill, T. J., 3rd, Thomsen, A. R., Tarrasch, J. T., Plouffe, B., Nguyen, A. H., Yang, F., et al. (2017). Distinct conformations of GPCR-beta-arrestin complexes mediate desensitization, signaling, and endocytosis. *Proceedings of the National Academy of Sciences of the United States of America*, *114*, 2562–2567.

Calebiro, D., Godbole, A., Lyga, S., & Lohse, M. J. (2015). Trafficking and function of GPCRs in the endosomal compartment. *Methods in Molecular Biology*, *1234*, 197–211.

Chaturvedi, M., Schilling, J., Beautrait, A., Bouvier, M., Benovic, J. L., & Shukla, A. K. (2018). Emerging paradigm of intracellular targeting of G protein-coupled receptors. *Trends in Biochemical Sciences*, *43*, 533–546.

Chen, G., Xie, R. G., Gao, Y. J., Xu, Z. Z., Zhao, L. X., Bang, S., et al. (2016). Beta-arrestin-2 regulates NMDA receptor function in spinal lamina II neurons and duration of persistent pain. *Nature Communications*, 7, 12531.

Costa-Neto, C. M., Parreiras, E. S. L. T., & Bouvier, M. (2016). A pluridimensional view of biased agonism. *Molecular Pharmacology*, *90*, 587–595.

Crudden, C., Song, D., Cismas, S., Trocme, E., Pasca, S., Calin, G. A., et al. (2019). Below the surface: IGF-1R therapeutic targeting and its endocytic journey. *Cell*, *8*, 1223.

DeWire, S. M., Ahn, S., Lefkowitz, R. J., & Shenoy, S. K. (2007). Beta-arrestins and cell signaling. *Annual Review of Physiology*, *69*, 483–510.

Draheim, K. M., Chen, H. B., Tao, Q., Moore, N., Roche, M., & Lyle, S. (2010). ARRDC3 suppresses breast cancer progression by negatively regulating integrin beta4. *Oncogene*, *29*, 5032–5047.

Endicott, J. A., & Ling, V. (1989). The biochemistry of P-glycoprotein-mediated multidrug resistance. *Annual Review of Biochemistry*, *58*, 137–171.

Feigin, M. E., Xue, B., Hammell, M. C., & Muthuswamy, S. K. (2014). G-protein-coupled receptor GPR161 is overexpressed in breast cancer and is a promoter of cell proliferation and invasion. *Proceedings of the National Academy of Sciences of the United States of America*, *111*, 4191–4196.

Fereshteha, M., Ito, T., Kovacs, J. J., Zhao, C., Kwon, H. Y., Tornini, V., et al. (2012). Beta-arrestin2 mediates the initiation and progression of myeloid leukemia. *Proceedings of the National Academy of Sciences of the United States of America*, *109*, 12532–12537.

Ge, L., Shenoy, S. K., Lefkowitz, R. J., & DeFea, K. (2004). Constitutive protease-activated receptor-2-mediated migration of MDA MB-231 breast cancer cells requires both beta-arrestin-1 and -2. *The Journal of Biological Chemistry*, *279*, 55419–55424.

Ghosh, E., Dwivedi, H., Baidya, M., Srivastava, A., Kumari, P., Stepniewski, T., et al. (2019). Conformational sensors and domain swapping reveal structural and functional differences between beta-arrestin isoforms. *Cell Reports*, *28*, 3287–3299.e3286.

Ghosh, E., Srivastava, A., Baidya, M., Kumari, P., Dwivedi, H., Nidhi, K., et al. (2017). A synthetic intrabody-based selective and generic inhibitor of GPCR endocytosis. *Nature Nanotechnology*, *12*, 1190–1198.

Goertzen, C. G., Dragan, M., Turley, E., Babwah, A. V., & Bhattacharya, M. (2016). KISS1R signaling promotes invadopodia formation in human breast cancer cell via beta-arrestin2/ERK. *Cellular Signalling*, *28*, 165–176.

Gold, L. (1995). The SELEX process: A surprising source of therapeutic and diagnostic compounds. *Harvey Lectures*, *91*, 47–57.

Goodman, O. B., Jr., Krupnick, J. G., Santini, F., Gurevich, V. V., Penn, R. B., Gagnon, A. W., et al. (1996). Beta-arrestin acts as a clathrin adaptor in endocytosis of the beta2-adrenergic receptor. *Nature*, *383*, 447–450.

Goodman, O. B., Jr., Krupnick, J. G., Santini, F., Gurevich, V. V., Penn, R. B., Gagnon, A. W., et al. (1998). Role of arrestins in G-protein-coupled receptor endocytosis. *Advances in Pharmacology*, *42*, 429–433.

Gurevich, V. V., & Gurevich, E. V. (2019). GPCR signaling regulation: The role of GRKs and arrestins. *Frontiers in Pharmacology*, *10*, 125.

Hara, M. R., Kovacs, J. J., Whalen, E. J., Rajagopal, S., Strachan, R. T., Grant, W., et al. (2011). A stress response pathway regulates DNA damage through beta2-adrenoreceptors and beta-arrestin-1. *Nature*, *477*, 349–353.

Jing, X., Zhang, H., Hu, J., Su, P., Zhang, W., Jia, M., et al. (2015). Beta-arrestin 2 is associated with multidrug resistance in breast cancer cells through regulating MDR1 gene expression. *International Journal of Clinical and Experimental Pathology*, *8*, 1354–1363.

Jonathan, W. K., Sun, J. P., Li, M., Pratico, E. D., Fereshteh, M. P., Ahrens, D. P., et al. (2014). Targeted disruption of beta-arrestin 2-mediated signaling pathways by aptamer chimeras leads to inhibition of leukemic cell growth. *PLoS One*, *9*, e93441.

Kang, D. S., Tian, X., & Benovic, J. L. (2014). Role of beta-arrestins and arrestin domain-containing proteins in G protein-coupled receptor trafficking. *Current Opinion in Cell Biology*, *27*, 63–71.
Kashihara, T., Nakada, T., Kojima, K., Takeshita, T., & Yamada, M. (2017). Angiotensin II activates CaV 1.2 Ca(2+) channels through beta-arrestin2 and casein kinase 2 in mouse immature cardiomyocytes. *The Journal of Physiology*, *595*, 4207–4225.
Kirby, H. R., Maguire, J. J., Colledge, W. H., & Davenport, A. P. (2010). International Union of Basic and Clinical Pharmacology. LXXVII. Kisspeptin receptor nomenclature, distribution, and function. *Pharmacological Reviews*, *62*, 565–578.
Koide, A., Wojcik, J., Gilbreth, R. N., Hoey, R. J., & Koide, S. (2012). Teaching an old scaffold new tricks: Monobodies constructed using alternative surfaces of the FN3 scaffold. *Journal of Molecular Biology*, *415*, 393–405.
Kotula, J. W., Pratico, E. D., Ming, X., Nakagawa, O., Juliano, R. L., & Sullenger, B. A. (2012). Aptamer-mediated delivery of splice-switching oligonucleotides to the nuclei of cancer cells. *Nucleic Acid Therapeutics*, *22*, 187–195.
Kovacs, J. J., Whalen, E. J., Liu, R., Xiao, K., Kim, J., Chen, M., et al. (2008). Beta-arrestin-mediated localization of smoothened to the primary cilium. *Science*, *320*, 1777–1781.
Kumari, P., Srivastava, A., Banerjee, R., Ghosh, E., Gupta, P., Ranjan, R., et al. (2016). Functional competence of a partially engaged GPCR-beta-arrestin complex. *Nature Communications*, 7, 13416.
Kumari, P., Srivastava, A., Ghosh, E., Ranjan, R., Dogra, S., Yadav, P. N., et al. (2017). Core engagement with beta-arrestin is dispensable for agonist-induced vasopressin receptor endocytosis and ERK activation. *Molecular Biology of the Cell*, *28*, 1003–1010.
Laporte, S. A., Oakley, R. H., Zhang, J., Holt, J. A., Ferguson, S. S., Caron, M. G., et al. (1999). The beta2-adrenergic receptor/betaarrestin complex recruits the clathrin adaptor AP-2 during endocytosis. *Proceedings of the National Academy of Sciences of the United States of America*, *96*, 3712–3717.
Lefkowitz, R. J., Rajagopal, K., & Whalen, E. J. (2006). New roles for beta-arrestins in cell signaling: Not just for seven-transmembrane receptors. *Molecular Cell*, *24*, 643–652.
Lefkowitz, R. J., & Shenoy, S. K. (2005). Transduction of receptor signals by beta-arrestins. *Science*, *308*, 512–517.
Li, T. T., Alemayehu, M., Aziziyeh, A. I., Pape, C., Pampillo, M., Postovit, L. M., et al. (2009). Beta-arrestin/Ral signaling regulates lysophosphatidic acid-mediated migration and invasion of human breast tumor cells. *Molecular Cancer Research*, 7, 1064–1077.
Liu, C. H., Gong, Z., Liang, Z. L., Liu, Z. X., Yang, F., Sun, Y. J., et al. (2017). Arrestin-biased AT1R agonism induces acute catecholamine secretion through TRPC3 coupling. *Nature Communications*, *8*, 14335.
Lobingier, B. T., & von Zastrow, M. (2019). When trafficking and signaling mix: How subcellular location shapes G protein-coupled receptor activation of heterotrimeric G proteins. *Traffic*, *20*, 130–136.
Lundgren, K., Tobin, N. P., Lehn, S., Stål, O., Rydén, L., Jirström, K., et al. (2011). Stromal expression of β-arrestin-1 predicts clinical outcome and tamoxifen response in breast cancer. *Journal of Molecular Diagnostics*, *13*(3), 340–351.
Martin, H. L., Smith, L., & Tomlinson, D. C. (2014). Multidrug-resistant breast cancer: Current perspectives. *Breast Cancer (Dove Med Press)*, *6*, 1–13.
Massara, M., Bonavita, O., Mantovani, A., Locati, M., & Bonecchi, R. (2016). Atypical chemokine receptors in cancer: Friends or foes? *Journal of Leukocyte Biology*, *99*, 927–933.
McMahon, C., Baier, A. S., Pascolutti, R., Wegrecki, M., Zheng, S., Ong, J. X., et al. (2018). Yeast surface display platform for rapid discovery of conformationally selective nanobodies. *Nature Structural & Molecular Biology*, *25*, 289–296.

Mollica Poeta, V., Massara, M., Capucetti, A., & Bonecchi, R. (2019). Chemokines and chemokine receptors: New targets for cancer immunotherapy. *Frontiers in Immunology, 10*, 379.

Nobles, K. N., Xiao, K., Ahn, S., Shukla, A. K., Lam, C. M., Rajagopal, S., et al. (2011). Distinct phosphorylation sites on the beta(2)-adrenergic receptor establish a barcode that encodes differential functions of beta-arrestin. *Science Signaling, 4*, ra51.

O'Brien, J., Hayder, H., Zayed, Y., & Peng, C. (2018). Overview of MicroRNA biogenesis, mechanisms of actions, and circulation. *Frontiers in Endocrinology (Lausanne), 9*, 402.

Parisis, N., Metodieva, G., & Metodiev, M. V. (2013). Pseudopodial and β-arrestin-interacting proteomes from migrating breast cancer cells upon PAR2 activation. *Journal of Proteomics, 80*, 91–106. https://doi.org/10.1016/j.jprot.2012.12.024.

Perry, N. A., Fialkowski, K. P., Kaoud, T. S., Kaya, A. I., Chen, A. L., Taliaferro, J. M., et al. (2019). Arrestin-3 interaction with maternal embryonic leucine-zipper kinase. *Cellular Signalling, 63*, 109366.

Peterson, Y. K., & Luttrell, L. M. (2017). The diverse roles of arrestin scaffolds in G protein-coupled receptor signaling. *Pharmacological Reviews, 69*, 256–297.

Que-Gewirth, N. S., & Sullenger, B. A. (2007). Gene therapy progress and prospects: RNA aptamers. *Gene Therapy, 14*, 283–291.

Reiter, E., & Lefkowitz, R. J. (2006). GRKs and beta-arrestins: Roles in receptor silencing, trafficking and signaling. *Trends in Endocrinology and Metabolism, 17*, 159–165.

Rezaul, K., Thumar, J. K., Lundgren, D. H., Eng, J. K., Claffey, K. P., Wilson, L., et al. (2010). Differential protein expression profiles in estrogen receptor-positive and -negative breast cancer tissues using label-free quantitative proteomics. *Genes Cancer, 1*(3), 251–271. https://doi.org/10.1177/1947601910365896.

Roy, I., Getschman, A. E., Volkman, B. F., & Dwinell, M. B. (2017). Exploiting agonist biased signaling of chemokines to target cancer. *Molecular Carcinogenesis, 56*, 804–813.

Sente, A., Peer, R., Srivastava, A., Baidya, M., Lesk, A. M., Balaji, S., et al. (2018). Molecular mechanism of modulating arrestin conformation by GPCR phosphorylation. *Nature Structural & Molecular Biology, 25*, 538–545.

Shenoy, S. K., & Lefkowitz, R. J. (2011). Beta-arrestin-mediated receptor trafficking and signal transduction. *Trends in Pharmacological Sciences, 32*, 521–533.

Shukla, A. K. (2014). Biasing GPCR signaling from inside. *Science Signaling, 7*, pe3.

Shukla, A. K., Kim, J., Ahn, S., Xiao, K., Shenoy, S. K., Liedtke, W., et al. (2010). Arresting a transient receptor potential (TRP) channel: Beta-arrestin 1 mediates ubiquitination and functional down-regulation of TRPV4. *The Journal of Biological Chemistry, 285*, 30115–30125.

Shukla, A. K., Manglik, A., Kruse, A. C., Xiao, K., Reis, R. I., Tseng, W. C., et al. (2013). Structure of active beta-arrestin-1 bound to a G-protein-coupled receptor phosphopeptide. *Nature, 497*, 137–141.

Shukla, A. K., Singh, G., & Ghosh, E. (2014). Emerging structural insights into biased GPCR signaling. *Trends in Biochemical Sciences, 39*, 594–602.

Shukla, A. K., Violin, J. D., Whalen, E. J., Gesty-Palmer, D., Shenoy, S. K., & Lefkowitz, R. J. (2008). Distinct conformational changes in beta-arrestin report biased agonism at seven-transmembrane receptors. *Proceedings of the National Academy of Sciences of the United States of America, 105*, 9988–9993.

Shukla, A. K., Westfield, G. H., Xiao, K., Reis, R. I., Huang, L. Y., Tripathi-Shukla, P., et al. (2014). Visualization of arrestin recruitment by a G-protein-coupled receptor. *Nature, 512*, 218–222.

Shukla, A. K., Xiao, K., & Lefkowitz, R. J. (2011). Emerging paradigms of beta-arrestin-dependent seven transmembrane receptor signaling. *Trends in Biochemical Sciences, 36*, 457–469.

Sobolesky, P. M., & Moussa, O. (2013). The role of beta-arrestins in cancer. *Progress in Molecular Biology and Translational Science, 118*, 395–411.

Son, D., Kim, Y., Lim, S., Kang, H. G., Kim, D. H., Park, J. W., et al. (2019). miR-374a-5p promotes tumor progression by targeting ARRB1 in triple negative breast cancer. *Cancer Letters, 454*, 224–233.

Soung, Y. H., Chung, H., Yan, C., Ju, J., & Chung, J. (2019). Arrestin domain containing 3 reverses epithelial to mesenchymal transition and chemo-resistance of TNBC cells by up-regulating expression of miR-200b. *Cell, 8*, 692.

Soung, Y. H., Ford, S., Yan, C., & Chung, J. (2018). The role of arrestin domain-containing 3 in regulating endocytic recycling and extracellular vesicle sorting of integrin beta4 in breast cancer. *Cancers (Basel), 10*, 507.

Soung, Y. H., Kashyap, T., Nguyen, T., Yadav, G., Chang, H., Landesman, Y., et al. (2017). Selective inhibitors of nuclear export (SINE) compounds block proliferation and migration of triple negative breast cancer cells by restoring expression of ARRDC3. *Oncotarget, 8*, 52935–52947.

Soung, Y. H., Pruitt, K., & Chung, J. (2014). Epigenetic silencing of ARRDC3 expression in basal-like breast cancer cells. *Scientific Reports, 4* 3846.

Srivastava, A., Gupta, B., Gupta, C., & Shukla, A. K. (2015). Emerging functional divergence of beta-arrestin isoforms in GPCR function. *Trends in Endocrinology and Metabolism, 26*, 628–642.

Thomsen, A. R. B., Plouffe, B., Cahill, T. J., 3rd, Shukla, A. K., Tarrasch, J. T., Dosey, A. M., et al. (2016). GPCR-G protein-beta-arrestin super-complex mediates sustained G protein signaling. *Cell, 166*, 907–919.

To, K. K. W., Fong, W., Tong, C. W. S., Wu, M., Yan, W., & Cho, W. C. S. (2020). Advances in the discovery of microRNA-based anticancer therapeutics: Latest tools and developments. *Expert Opinion on Drug Discovery, 15*, 63–83.

Tocci, P., Rosano, L., & Bagnato, A. (2019). Targeting endothelin-1 receptor/beta-arrestin-1 Axis in ovarian cancer: From basic research to a therapeutic approach. *Frontiers in Endocrinology (Lausanne), 10*, 609.

Tsvetanova, N. G., Irannejad, R., & von Zastrow, M. (2015). G protein-coupled receptor (GPCR) signaling via heterotrimeric G proteins from endosomes. *The Journal of Biological Chemistry, 290*, 6689–6696.

Vilardaga, J. P., Jean-Alphonse, F. G., & Gardella, T. J. (2014). Endosomal generation of cAMP in GPCR signaling. *Nature Chemical Biology, 10*, 700–706.

Wang, C., Chen, W., & Shen, J. (2018). CXCR7 targeting and its major disease relevance. *Frontiers in Pharmacology, 9*, 641.

Wu, J. X., Shan, F. X., Zheng, J. N., & Pei, D. S. (2014). Beta-arrestin promotes c-Jun N-terminal kinase mediated apoptosis via a GABA(B)R beta-arrestin JNK signaling module. *Asian Pacific Journal of Cancer Prevention, 15*, 1041–1046.

Xiao, K., McClatchy, D. B., Shukla, A. K., Zhao, Y., Chen, M., Shenoy, S. K., et al. (2007). Functional specialization of beta-arrestin interactions revealed by proteomic analysis. *Proceedings of the National Academy of Sciences of the United States of America, 104*, 12011–12016.

Xiao, K., Sun, J., Kim, J., Rajagopal, S., Zhai, B., Villen, J., et al. (2010). Global phosphorylation analysis of beta-arrestin-mediated signaling downstream of a seven transmembrane receptor (7TMR). *Proceedings of the National Academy of Sciences of the United States of America, 107*, 15299–15304.

Yamaguchi, H. (2012). Pathological roles of invadopodia in cancer invasion and metastasis. *European Journal of Cell Biology, 91*, 902–907.

Yang, Z., Yang, F., Zhang, D., Liu, Z., Lin, A., Liu, C., et al. (2017). Phosphorylation of G protein-coupled receptors: From the barcode hypothesis to the flute model. *Molecular Pharmacology, 92*, 201–210.

Yin, W., Li, Z., Jin, M., Yin, Y. L., de Waal, P. W., Pal, K., et al. (2019). A complex structure of arrestin-2 bound to a G protein-coupled receptor. *Cell Research*, *29*, 971–983.

Zajac, M., Law, J., Cvetkovic, D. D., Pampillo, M., McColl, L., Pape, C., et al. (2011). GPR54 (KISS1R) transactivates EGFR to promote breast cancer cell invasiveness. *PLoS One*, *6*(6), e21599. https://doi.org/10.1371/journal.pone.0021599.

Zhao, M., Zhou, G., Zhang, Y., Chen, T., Sun, X., Stuart, C., et al. (2009). Beta-arrestin2 inhibits opioid-induced breast cancer cell death through Akt and caspase-8 pathways. *Neoplasma*, *56*, 108–113.

Zheng, Y., Lin, Z. Y., Xie, J. J., Jiang, F. N., Chen, C. J., Li, J. X., et al. (2017). ARRDC3 inhibits the progression of human prostate cancer through ARRDC3-ITGbeta4 pathway. *Current Molecular Medicine*, *17*, 221–229.

Kolibri 360 GmbH
Ferdinand-Jühlke-Straße 7
99095 Erfurt, Deutschland
produktsicherheit@kolibri360.de